KU-359-821

MULTILINGUAL DICTIONARY OF CONCRETE

MULTILINGUAL DICTIONARY OF CONCRETE

A compilation of terms in English, French, German, Spanish, Dutch and Russian

compiled under the auspices of

FÉDÉRATION INTERNATIONALE DE LA PRÉCONTRAINTE (FIP)

Wexham Springs, Slough, Bucks., England

ELSEVIER SCIENTIFIC PUBLISHING COMPANY

AMSTERDAM / OXFORD / NEW YORK

1976

ELSEVIER SCIENTIFIC PUBLISHING COMPANY
335 Jan van Galenstraat
P.O. Box 211, Amsterdam, The Netherlands

Distributors for the United States and Canada:

ELSEVIER/NORTH-HOLLAND INC.
52, Vanderbilt Avenue
New York, N.Y. 10017

ISBN: 0-444-41237-9

Printed in The Netherlands

Introduction

International exchanges of thoughts and ideas are too often hampered by language problems in all disciplines. This is especially true in the building profession and in the world of concrete construction.

FIP is most grateful to the small group of engineers from the Netherlands who undertook the task of preparing a list of concrete terms in Dutch and, with the help of a translator, in English also. From this initiative within the Dutch member-group STUVO cooperation was subsequently obtained from other FIP national groups in France, the Federal Republic of Germany, U.S.S.R., Spain, Great Britain and the United States of America for its completion. Finally the Elsevier Scientific Publishing Company of Amsterdam agreed to produce the language dictionary of concrete for FIP.

This project was initiated and largely completed under the general direction of the immediate past president of FIP, Dr. G.F. Janssonius, whose persistence in the face of many difficulties has led now to this significant contribution to better communication and understanding.

I am extremely pleased to introduce this very useful publication to engineers working in the field of concrete throughout the world, and extend a warm vote of thanks to all those who contributed to the successful finalization of this project.

Ben C. Gerwick Jr.,

President of FIP

Preface

The Fédération Internationale de la Précontrainte (FIP) and its Dutch membergroup STUVO were jointly concerned with the original idea for a Multilingual Dictionary of Concrete to help those engaged in the construction industry, but its scope covers reinforced concrete as well as prestressed concrete terms since they are so often used together.

The STUVO appointed a special Commission whose names appear below to compile the basic list of some 1500 words, aiming as far as possible to include items which were new in the field of concrete construction.

The present may be seen as a successor, but not a second edition, of a previous FIP publication which appeared some fourteen years ago after the Rome/Naples Congress. The lay-out arrangement and method of use are very different.

The Commission:

Ir. M.R. Robaard, Chairman
Ing. M.G.P. Nelissen, Secretary
Ir. P.H. Jansma
Ir. W. Stevelink
Ir. R. Swart
Ir. R.H. van der Wateren

BASIC·TABLE

A

0001 abrasion resistance
f résistance à l'abrasion
d Abriebfestigkeit
e resistencia a la abrasión
n slijtweerstand

0002 absorption
f absorption
d Absorption
e absorción
n absorptie

0003 absorption of water
f absorption d'eau
d Wasseraufnahme
e absorción de agua
n wateropname

0004 abutment
f culée
d Widerlager
e estribo
n landhoofd

0005 accelerate *v*
f accélérer
d beschleunigen
e acelerar
n versnellen

* **account** → 0117

0006 acidity (pH)
f acidité (pH)
d Säuregrad (pH)
e acidez (pH)
n zuurgraad (pH)

0007 acid-resistant
f résistant aux acides
d säurebeständig
e resistente a los ácidos
n zuurbestendig

0008 acoustic testing
f essai acoustique
d Schallprüfung
e ensayo acústico
n akoestisch onderzoek

0009 add *v*
f ajouter
d hinzufügen
e añadir; adicionar
n toevoegen

* **adhesive** → 0129

0010 adhesive tape
f ruban adhésif
d Klebeband
e tela adhesiva
n kleefband

0011 adjust *v*
f ajuster; régler
d einstellen; regulieren
e ajustar
n stellen

0012 adjusting screw
f vis de réglage
d Stellschraube
e tornillo de ajuste; tornillo de regulación
n stelschroef

0013 admixture
f adjuvant
d Zusatzmittel
e material auxiliar
n hulpstof; toevoeging

0014 age
f âge
d Alter
e edad
n ouderdom

0015 agent
f chef de chantier
d Bauleiter
e agente
n uitvoerder

0016 aggregate
f agrégat; granulat
d Zuschlag(stoff)
e árido
n toeslagmateriaal

0017 aggressive
f agressif
d aggressiv
e agresivo
n agressief

0018 agitator
f agitateur
d Rührwerk
e agitador
n agitator

0019 air
f air
d Luft
e aire
n lucht

0020 air bubble
f bulle d'air
d Luftblase
e burbuja de aire
n luchtbel

0021 air-conditioned room
f salle climatisée
d klimatisierter Raum
e habitación con aire acondicionado
n klimaatkamer

0022 air content
f teneur en air
d Luftgehalt
e contenido de aire
n luchtgehalte

0023 air-entraining agent
f agent entraîneur d'air
d Luftporenbildner; Belüftungsmittel
e agente aireante
n luchtbelvormer

0024 air meter
f appareil à mesurer l'entraînement d'air
d Luftmessgerät
e medidor de aire
n luchtmeter

0025 air-tight concrete
f béton imperméable à l'air
d luftundurchlässiger Beton
e hormigón estanco al aire; hormigón impermeable al aire
n luchtdicht beton

0026 air voids; pores
f pores; vides
d Luftporen; Hohlräume
e poros de aire
n luchtporiën

0027 alkaline
f alcalin
d alkalisch
e alcalino
n alkalisch

0028 alloy
f alliage
d Legierung
e aleación
n legering

0029 alternating load
f charge alternée
d Wechselbelastung
e carga alternada
n wisselende belasting

0030 alternating stress
f contrainte alternée
d Wechselbeanspruchung
e tensión alterna
n wisselspanning

0031 aluminous cement; high-alumina cement
f ciment alumineux
d Tonerdezement
e cemento aluminoso
n aluminiumcement

0032 amplitude
f amplitude
d Amplitude
e amplitud
n amplitude

0033 anchor
f ancre
d Anker
e ancla
n anker

0034 anchor bearing plate
f plaque d'ancrage
d Ankerplatte
e placa de apoyo del anclaje
n ankerplaat

0035 anchor cone
f cône d'ancrage
d Verankerungskonus
e cono de anclaje
n ankerconus

0036 anchor cup
f cloche d'ancrage
d Verankerungsglocke
e trompeta de anclage
n ankerklok

0037 anchor female cone
f cône d'ancrage femelle
d äusserer Verankerungskonus
e cono hembra de anclaje
n ankerconus

0038 anchor male cone
f cône d'ancrage mâle
d innerer Verankerungskonus
e cono macho de anclaje
n ankerprop

0039 anchor nut
f écrou d'ancrage
d Verankerungsmutter
e tuerca de anclaje
n ankermoer

0040 anchor plate
f plaque d'ancrage
d Verankerungsplatte; Ankerplatte
e placa de anclaje
n ankerplaat

0041 anchor wedge
f clavette d'ancrage
d Verankerungskeil; Ankerkeil
e cuña de anclaje
n ankerwig

0042 anchorage
f ancrage
d Verankerung
e anclaje
n verankering

0043 anchorage length
f longueur d'ancrage
d Verankerungslänge
e longitud de anclaje
n verankeringslengte

0044 anchorage zone
f zone d'ancrage
d Verankerungszone
e zona de anclaje
n verankeringszone

0045 angle (in degrees)
f angle (en degrés)
d Winkel (in Graden)
e ángulo (en grados)
n hoek (in graden)

0046 angle of friction
f angle de frottement
d Reibungswinkel
e ángulo de rozamiento
n wrijvingshoek

0047 angle of twist
f angle de torsion
d Verdrehungswinkel
e ángulo de giro
n wringingshoek

0048 angular rotation
f rotation angulaire
d Winkeldrehung
e rotación angular
n hoekverdraaiing

0049 anisotropic
f anisotrope
d anisotrop
e anisótropo
n anisotroop

0050 anneal *v*
f recuire
d anlassen; tempern; glühen
e recocer
n uitgloeien

0051 annealed wire
f fil recuit
d geglühter Draht
e alambre recocido
n uitgegloeid draad

0052 apparent specific gravity
f masse volumique des grains
d Kornrohdichte
e densidad volumétrica de la partícula
n korrelvolumegewicht

0053 approach span
f travée d'approche
d Landfeld
e tramo de acceso
n aanbrug

0054 aqueduct
f aqueduc
d Aquedukt
e acueducto
n aquaduct

0055 arch
f arc; voûte
d Bogen
e arco
n boog

0056 arch bridge
f pont en arc
d Bogenbrücke
e puente en arco
n boogbrug

0057 architect
f architecte
d Architekt
e arquitecto
n architect

0058 arm (of a couple)
f bras de levier (d'un couple)
d Hebelarm (eines Kräftepaares)
e brazo (de un par)
n arm (van een koppel)

0059 artificial ageing
f vieillissement artificiel
d künstliches Altern
e envejecimiento artificial
n kunstmatige veroudering

0060 artificial foam
f mousse artificielle
d Kunstschaum
e espuma artificial
n kunstschuim

0061 assemble *v*
f assembler
d zusammenbauen
e ensamblar; unir
n verbinden; samenvoegen

0062 asymptote
f asymptote
d Asymptote
e asíntota
n asymptoot

0063 attach *v*
f attacher
d befestigen; anfügen
e enganchar; conectar
n aanhechten

0064 autoclave
f autoclave
d Autoklav
e autoclave
n autoclaaf

0065 auxiliary structure
f structure auxiliaire
d Hilfskonstruktion
e estructura auxiliar
n hulpconstructie

0066 avalanche load
f charge d'avalanche
d Lawinenbelastung
e carga de alud
n lawinebelasting

0067 average; mean
f moyenne
d Durchschnitt; Mittelwert
e promedio
n gemiddelde

0068 average stress
f contrainte moyenne
d durchschnittliche Beanspruchung
e tensión media
n gemiddelde spanning

0069 award (of a contract); **letting** (of a contract)
f passation (du marché)
d Auftragserteilung; Vergabe (einer Arbeit)
e adjudicación (de una obra)
n gunning (van een werk)

0070 axial
f axial
d axial; mittig
e axíl
n axiaal

0071 axial load
f charge centrale
d zentrische Belastung
e carga central
n centrische belasting

0072 axial stress
f contrainte axiale
d mittige Beanspruchung
e tensión axíl
n axiaalspanning

0073 axis (of a diagram)
f axe (d'un diagramme)
d Achse (eines Diagramms)
e eje (de un diagrama)
n as (van een figuur)

0074 axle load
f charge d'essieu
d Achslast
e carga por eje
n asbelasting

B

0075 balance; scale
f balance
d Waage
e balanza
n weegschaal

0076 balance beam
f balancier
d Schwinge
e balancín
n evenaar

0077 balcony slab
f dalle de balcon
d Balkonplatte
e placa balcón
n balkonplaat

0078 ball bearing
f roulement à billes
d Kugellager
e cojinete de bolas
n kogellager

0079 ball joint
f joint à billes
d Kugelgelenk
e articulación esférica; rótula esférica
n kogelgewricht

0080 ball mill
f broyeur à boulets
d Kugelmühle
e molino de bolas
n kogelmolen

0081 bar
f barre
d Stab
e barra
n staaf

0082 bar mesh
f treillis
d Bewehrungsmatte
e malla metálica
n kruisnet

0083 bar section
f section d'une barre
d Stabquerschnitt
e sección de una barra
n staafdoorsnede

0084 bar spacing
f espacement de barres
d Stababstand
e distancia entre barras; separación entre barras
n staafafstand

0085 barrage
f barrage
d Stauanlage; Wehr
e presa; dique
n stuwdam

0086 barrel-vault shell
f coque-voûte
d tonnenschale
e presa bóveda
n tonschaal

0087 base (of a foundation)
f base (d'une fondation)
d Sohle (eines fundaments)
e base (de una cimentación)
n grondslag (van een fundering)

0088 basement; cellar
f sous-sol; cave
d Keller(geschoss)
e basamento; bodega; cueva
n kelder

0089 basis (of a calculation)
f base (d'u calcul)
d Grundlage (einer Berechnung)
e bases (de cálculo)
n grondslag (van een berekening)

0090 batching
f dosage
d Dosierung; Abmessung
e amasada
n dosering

* **batter pile** → 0934

0091 bay (of a building)
f nef (d'un bâtiment)
d Feld; Hallenschiff
e crujía (de un edificio)
n travee

0092 beam (girder)
f poutre
d Balken (Träger)
e viga (jácena; carrera)
n balk (ligger)

0093 beam grid; beam grillage
f grillage de poutres
d Trägerrost
e emparrillado de vigas
n balkrooster

* **beam grillage** → 0093

0094 beam section
f section d'une poutre
d Trägerquerschnitt
e sección de la viga
n balkprofiel

0095 bearing
f appui
d Auflager
e apoyo
n oplegging

0096 bearing block
f bloc d'appui
d Lagerblock
e bloque de apoyo
n oplegblok

0097 bearing length
f longueur d'appui
d Auflagerlänge
e longitud de apoyo; entrega
n opleglengte

0098 bearing pressure
f pression d'appui
d Auflagerdruck
e presión de apoyo
n oplegdruk

0099 bearing reaction
f réaction d'appui
d Auflagerreaktion
e reacción de apoyo
n oplegreactie

0100 belt conveyor
f transporteur à courroie
d Förderband; Bandförderer
e cinta transportadora
n transportband

0101 bend down *v*
f plier vers le bas
d niederbiegen
e doblar hacia abajo
n neerbuigen

0102 bend up *v* (of cables)
f relever (des câbles)
d aufbiegen (von Kabeln)
e levantar (de cables)
n opbuigen (van kabels)

0103 bending and reinforcement assembly shop
f atelier de ferraillage
d Biege- und Flechtbetrieb
e taller de ferralla
n vlechtcentrale

0104 bending bench
f banc de pliage
d Biegebank
e banco de doblado
n buigbank

0105 bending machine
f plieuse
d Biegemaschine
e máquina de doblado
n buigmachine

0106 bending moment
f moment fléchissant
d Biegemoment
e momento flector
n buigend moment

0107 bending moment diagram
f diagramme de moments
d Momentendiagramm
e diagrama de momentos
n momentenvlak

0108 bending schedule
f tableau de pliage
d Biegeplan
e lista de doblado; plantilla de doblado
n buigstaat

0109 bending test
f essai de flexion
d Biegeversuch
e ensayo de doblado
n buigproef

0110 bent
f palée
d Joch
e cabellete
n juk

0111 bent-down bar
f barre pliée vers le bas
d niedergebogener Bewehrungsstab
e barra doblada hacia abajo
n zakstaaf

0112 bent-up bar
f barre relevée
d aufgebogener Bewehrungsstab
e barra levantada
n opgebogen staaf

0113 bent-up reinforcement
f armature relevée
d Bewehrung aus aufgebogenen Stäben
e armadura levantada
n opgebogen wapening

0114 bevel; chamfer
f chanfrein
d Abschränkung; Abfasung; Schrägkante
e bisel; chaflán
n afschuining; vellingkant

0115 biaxial stress
f contrainte biaxiale
d zweiachsige Beanspruchung
e tensión biaxíl
n biaxiaalspanning

0116 bid; tender
f offre
d Angebot
e propuesta; oferta; licitación
n offerte; inschrijving (voor een aanbesteding)

0117 bill; account
f facture
d Rechnung
e factura; cuenta
n rekening

* **bin** → 1080

0118 bind *v* (with reinforcement)
f fretter
d umschnüren
e enrollar; zunchar
n omwikkelen

0119 binding agent
f liant
d Bindemittel
e aglomerante
n bindmiddel

0120 binding reinforcement
f armature de frettage
d Umschnürungsbewehrung
e zuncho; armadura de zunchado
n omwikkelingswapening

0121 bitumen
f bitume
d Bitumen
e betún
n bitumen

0122 blast-furnace cement
f ciment de haut-fourneau
d Hochofenzement
e cemento de horno alto; cemento siderúrgico
n hoogovencement

0123 blast-furnace slag
f laitier de haut-fourneau
d Hochofenschlacke
e escoria
n hoogovenslak

0124 bleed *v*; **vent** *v*
f ventiler
d entlüften
e exudar
n ontluchten

0125 bleeding; water segregation
f ressuage
d Wasserabsonderung
e exudación; segregación de agua
n waterafscheiding

0126 blinding; sealing coat
f béton de scellement
d Unterbeton
e ocultación
n werkvloer

0127 bolt
f boulon
d Schraubenbolzen
e bulón; perno; clavija; pasador
n bout

0128 bond stress
f contrainte d'adhérence
d Haftspannung; Verbundspannung
e tensión de adherencia
n aanhechtspanning

0129 bonding agent; adhesive
f matière liante
d Kitt
e agente adherente; adhesivo
n kit (lijm)

0130 bottom bar
f armature inférieure
d unterer Stab
e barra inferior
n onderstaaf

0131 bottom fibre
f fibre inférieure
d untere (Rand)faser
e fibra inferior
n onderste vezel

0132 bottom mesh
f treillis soudé inférieur
d unteres Bewehrungsnetz
e malla inferior
n ondernet

0133 boundary condition
f condition aux limites
d Randbedingung
e condición de borde
n randvoorwaarde

0134 box-frame construction
f construction en caisson
d Kastenbauweise
e construcción en cajón
n doosconstructie

0135 box girder
f poutre caisson
d Kastenträger
e viga-cajón
n kokerligger

0136 brace
f étrésillon
d Strebe
e viga apuntalada
n schoorstaaf

0137 bracket
f console
d Konsole
e ménsula; cartela; escuadra
n console

* **breadth** → 1387

0138 bridge
f pont
d Brücke
e puente
n brug

0139 bridge deck
f tablier (de pont)
d Brückentafel; Fahrbahntafel
e tablero de puente
n brugdek; rijdek

0140 bridge girder
f poutre de pont
d Brückenträger
e viga de puente
n brugbalk

0141 bridge pier
f pile de pont
d Brückenpfeiler
e pila de puente
n brugpijler

0142 broken
f cassé; rompu
d gebrochen
e roto; quebrado; machacado
n gebroken

0143 buckling (of a column)
f flambage (d'une colonne)
d Knicken (einer Säule)
e pandeo (de un soporte)
n knik (van een kolom)

0144 buckling (of plates)
f voilement (des plaques)
d Beulen (von Platten)
e alabeo (de placas); pandeo (de

placas)
n knikken (plooien)

0145 buckling coefficient
f coefficient de flambage
d Knickbeiwert; Beulbeiwert
e coeficiente de pandeo
n knikfactor

0146 buckling strength
f résistance au flambement
d Knickfestigkeit; Beulfestigkeit
e resistencia al pandeo
n kniksterkte

0147 buckling stress
f contrainte de flambement
d Knickspannung; Beulspannung
e tensión de pandeo
n knikspanning

0148 building
f bâtiment; construction
d Gebäude; Bauwerk
e edificio
n gebouw; bouwwerk

0149 building authorities
f autorité surveillant des constructions
d Baubehörden
e autoridades oficiales que regulan la construcción
n bouwpolitie

0150 building owner; client; promotor
f maître de l'ouvrage
d Bauherr; Auftraggeber
e propietario; cliente; dueño
n bouwheer

0151 building regulations
f règlement de construction
d Baubestimmungen
e normas de edificación
n bouwvoorschriften

0152 building supervision
f surveillance des travaux
d Bauaufsicht
e supervisión de la obra
n bouwtoezicht

0153 bulb steel
f fer à bourrelet
d Wulststahl
e acero de bulbo
n bulbstaal

0154 bulk cement
f ciment en vrac
d loser Zement; Silozement
e cemento a granel
n cement in bulk

0155 bulk density
f poids volumique
d Raumgewicht
e densidad volumétrica
n volumegewicht

0156 burn *v* the wires
f couper de fils au chalumeau
d durchbrennen der Drähte
e cortar de alambre con soplete
n doorbranden van draden

0157 bush hammering
f bouchardage
d Scharrieren
e abucardamiento
n boucharderen

0158 butt joint
f assemblage bout à bout
d stumpfer Stoss
e junta a tope
n stuik

0159 butt strap; cover plate
f couvre-joint
d Decklasche
e listón; rastrel
n koppelplaat

0160 button-head anchorage
f ancrage par boutonnage
d Stauchkopfverankerung
e anclaje de cabezas recalcadas
n knopverankering

0161 buttress
f contrefort
d Strebepfeiler
e machón; contrafuerte
n steunbeer

C

0162 cable
f câble
d Drahtseil; Seil; Kabel
e cable
n kabel

0163 cable clamp
f serre-câble
d Kabelklemme
e grapa de cable; mordaza de cable; prisionero de cable
n kabelklem

0164 cable duct
f gaine de câble
d Spannkanal
e conducto para alojar el cable
n kabelkanaal

0165 cable grip
f chaussette de traction
d Kabelziehstrumpf
e tensor de cable
n kabelvlieter

0166 cable sheath
f gaine de câble
d Kabelhüllrohr
e vaina de cable; funda de cable
n kabelomhulling

0167 cable-stayed bridge
f pont à haubans
d seilverspannte Brücke
e puente atirantado
n tuibrug

0168 cadmium plating
f plaquage de cadmium
d kadmieren
e cadmiando
n cadmeren

0169 caisson
f caisson
d Senkkasten; Caisson
e cajón de cimentación
n caisson

0170 calculate *v*; **compute** *v*
f calculer
d berechnen
e calcular
n berekenen

0171 calibrate *v*
f calibrer; étalonner
d eichen
e calibrar; contrastar
n ijken; kalibreren

0172 calibration value
f valeur d'étalonnage
d Eichwert
e valor de calibración
n ijkwaarde

0173 camber
f cambrure
d Überhöhung
e contraflecha
n zeeg; porring

0174 cantilever
f travée en porte-à-faux; travée en cantilever
d Kragarm
e ménsula; voladizo
n uitkraging; overstek; console

0175 cantilever bridge
f pont cantilever
d Kragbrücke; Auslegerbrücke
e puente ménsula; puente en voladizo
n kraagliggerbrug

0176 cantilever girder
f poutre en porte-à-faux
d Kragbalken(träger)
e viga ménsula; viga en voladizo
n kraagligger

0177 cantilevering
f en porte-à-faux
d freitragend
e en voladizo
n vrijdragend (van een vloer)

0178 cantilevering construction (of bridges)
f construction (d'un pont) en porte-à-faux
d freier Vorbau (von Brücken)
e construcción en voladizo (de puentes)
n vrije uitbouw (van bruggen)

0179 cap cable
f câble chapeau
d Hutkabel
e cable sombrero; armadura de continuidad
n hoedkabel

0180 capacity
f capacité
d Inhalt; Fassungsvermögen
e capacidad
n inhoud

0181 capillary
f capillaire
d kapillar
e capilar
n capillair

0182 cardboard
f carton
d Pappe
e cartón
n karton

0183 carriageway
f chaussée
d Fahrbahn
e calzada
n rijbaan

0184 cast in *v*
f enrober
d einbetonieren
e moldear en; hormigonar en
n instorten

0185 cast iron
f fonte
d Gusseisen
e fundición
n gietijzer

0186 cast steel
f acier moulé
d Gusstahl
e acero moldeado; acero colado
n gietstaal

* **casting** → 0283

0187 cavity
f cavité
d Hohlraum
e cavidad
n holle ruimte

* **cellar** → 0088

0188 cellular concrete
f béton cellulaire
d Zellenbeton
e hormigón celular
n cellenbeton

0189 cement
f ciment
d Zement
e cemento
n cement

* **cement bin** → 0197

0190 cement chemistry
f chimie du ciment
d Zementchemie
e química del cemento
n cementchemie

0191 cement clinker
f clinker
d Zementklinker
e clinker de cemento
n cementklinker

0192 cement content
f teneur en ciment
d Zementgehalt
e contenido de cemento; dosificación de cemento
n cementgehalte

0193 cement gel
f gel de ciment
d Zementgel
e gel de cemento
n cementgel

0194 cement in bulk
f ciment en vrac
d loser Zement; unverpackter Zement
e cemento a granel
n cement in bulk

0195 cement mortar
f mortier de ciment
d Zementmörtel

e mortero de cemento
n cementmortel

0196 cement paste
f pâte de ciment
d Zementleim
e pasta de cemento
n cementlijm; cementpasta

0197 cement silo; cement bin
f silo à ciment
d Zementsilo
e silo de cemento
n cementsilo

0198 cement skin
f pellicule de ciment
d Zementhaut
e película de cemento; lechada superficial de cemento
n cementhuid

0199 cement slurry
f coulis de ciment
d Zementschlämme
e pasta de cemento; lechada de cemento
n cementslib

0200 centre *v*
f centrer
d zentrieren
e centrar
n centreren

0201 centre line
f ligne médiane; axe
d Symmetrieachse
e línea central; línea media
n as; hartlijn

0202 centre of gravity
f centre de gravité
d Schwerpunkt
e centro de gravedad
n zwaartepunt

0203 centre of pressure
f centre de poussée; centre de pression
d Druckzentrum
e centro de presión
n drukpunt

0204 centre-to-centre distance
f distance entre axes
d Mittenabstand
e distancia entre centros
n hart op hart afstand

0205 centrifugally cast concrete
f béton centrifugé
d Schleuderbeton
e hormigón centrifugado
n gecentrifugeerd beton

0206 centrifuge *v*
f centrifuger
d zentrifugieren
e centrifugar
n centrifugeren

0207 centring pin
f broche de centrage
d Zentrierstift
e pasador de centrado; clavija de centrado
n centerpen

0208 centroidal axis
f axe de gravité
d Schwerachse
e mediana; directriz
n zwaartelijn

0209 certificate (of approval) (agrément)
f certificat (d'agrément)
d Bescheid (Zulassungs-)
e certificado (de aprobación)
n certificaat (van goedkeuring)

* **chamfer** → 0114

0210 characteristic
f caractéristique
d charakteristisch
e característico
n karakteristiek

0211 characteristic diameter
f diamètre caractéristique
d Kenndurchmesser
e diámetro característico
n kendiameter

0212 characteristic strength
f résistance caractéristique
d Kennfestigkeit

e resistencia característica
n karakteristieke sterkte

0213 checking test
f essai de contrôle
d Kontrollversuch
e ensayo de comprobación
n controleproef

0214 chemical testing
f essai chimique
d chemische Untersuchung
e ensayo químico
n chemisch onderzoek

0215 chemically resistant
f résistant aux agents chimiques
d chemisch beständig
e resistente a los agentes químicos
n chemisch-bestendig

0216 chief engineer
f ingénieur en chef
d Oberingenieur
e ingeniero jefe
n hoofdingenieur

0217 chippings
f concassé
d Splitt
e almendrilla; grava fina
n gebroken natuursteen; split; steenslag

0218 chop *v*
f couper
d hacken
e rajar; cortar; hender
n hakken

0219 chop off *v*
f découper
d abhacken
e tronchar; cortar
n afhakken

0220 chuted concrete
f béton fluide
d Gussbeton; Rinnenbeton
e hormigón vertido por canaleta
n gietbeton

0221 circumference; perimeter
f circonférence; périmètre
d Umfang
e circunferencia; perímetro
n omtrek

0222 circumferential prestress(ing)
f précontrainte circonférentielle
d Ringvorspannung
e pretensado circunferencial; pretensado anular
n ringvoorspanning

0223 clamp; clip
f agrafe
d Klammer
e grapa; pinza
n klem; klamp

0224 clay
f argile
d Ton
e arcilla
n klei

0225 clear width
f largeur libre
d lichte Breite
e anchura libre
n dagwijdte

0226 clearance
f gabarit
d lichter Raum
e holgura
n tussenruimte; vrije ruimte

0227 clearance gauge
f jauge de gabarit
d Lichtraumprofil
e gálibo
n profiel van vrije ruimte

* **client** → 0150

* **clip** → 0223

0228 coarse aggregate
f gros agrégat
d grobes Zuschlagsmaterial
e árido grueso
n grof toeslagmateriaal

0229 coarse gravel
f gros gravier
d Grobkies

e grava gruesa
n grof grind

0230 coat *v* **with cement slurry**
f enduire de coulis de ciment
d mit Zementschlämme anstreichen
e revestir con pasta o lechada de cemento
n aanbranden

0231 coefficient
f coefficient
d Koeffizient; Beiwert
e coeficiente
n coëfficient

0232 coefficient of expansion
f coefficient d'expansion; coefficient de dilatation
d Ausdehnungskoeffizient
e coeficiente de dilatación
n uitzettingscoëfficient

0233 coefficient of friction
f coefficient de frottement
d Reibungskoeffizient; Reibungsbeiwert
e coeficiente de rozamiento
n wrijvingscoëfficient

0234 coefficient of variation
f coefficient de variation
d Variationskoeffizient
e coeficiente de variación
n variatiecoëfficient; spreidingscoëfficient

0235 colcrete concrete
f béton colcrete
d Colcrete-Beton
e hormigón "colcrete"
n colcrete beton

0236 cold-drawn steel
f acier tréfilé à froid
d kaltgezogener Stahl
e acero estirado en frío
n koudgetrokken staal

0237 cold-drawn wire
f fil tréfilé à froid
d kaltgezogener Draht
e alambre estirado en frío; alambre trefilado
n koudgetrokken draad

0238 cold-worked steel
f acier écroui à froid
d kaltverformter Stahl
e acero deformado en frío
n koudvervormd staal

0239 collaborate *v*
f collaborer
d zusammenarbeiten
e colaborar
n samenwerken

0240 collapse *v*
f s'effondrer
d einstürzen
e colapsar; derrumbar; hundir
n instorten; bezwijken

0241 collapse load
f charge de ruine
d Bruchlast; Einsturzlast
e carga de hundimiento
n bezwijklast; bezwijkbelasting

0242 column
f colonne
d Stütze; Säule
e soporte; columna; pilar
n kolom

0243 column axis
f axe d'une colonne
d Stützenachse
e eje del soporte
n kolomas

0244 column base
f base de colonne
d Stützenfuss
e base del soporte
n kolomvoet

0245 column footing
f pied d'une colonne
d Stützenfundament
e zapata del soporte
n kolomvoet (fundament)

0246 column head
f chapiteau
d Stützenkopf
e cabeza del soporte
n kolomkop

0247 column strip
f bande d'appui
d Gurtstreifen
e banda de soportes
n kolomstrook

0248 column with binders or ties
f poteau armé de cadres
d verbügelte Säule
e soporte zunchado
n bebeugelde kolom

* **column with helical binding**
→ 1133

0249 commercial quality
f qualité commerciale
d Handelsqualität
e calidad comercial
n handelskwaliteit

0250 compact *v*
f compacter
d verdichten
e compactar; consolidar
n verdichten

0251 compaction index (according to Walz)
f degré de compacité (d'après Walz)
d Verdichtungsgrad (nach Walz)
e grado de compactación (según Walz)
n verdichtingsmaat (volgens Walz)

0252 complementary
f complémentaire
d ergänzend
e complementario
n complementair

0253 component
f composant
d Bestandteil
e componente
n onderdeel

0254 compose *v*
f composer
d zusammensetzen
e componer; formar
n samenstellen

0255 composite column
f colonne composite
d Verbundstütze
e soporte compuesto
n samengestelde kolom

0256 compress *v*
f comprimer
d komprimieren
e comprimir
n samendrukken

0257 compressed air
f air comprimé
d Pressluft
e aire comprimido
n samengeperste lucht; perslucht

0258 compression; pressure
f compression
d Druck
e compresión
n druk

0259 compression flange
f membrure comprimée
d Druckgurt
e ala comprimida; cabeza de compresión
n drukflens

0260 compression reinforcement
f armature comprimée
d Druckbewehrung
e armadura de compresión
n drukwapening

0261 compression-testing machine
f presse d'essai
d Prüfpresse
e máquina para el ensayo de compresión
n drukpers

0262 compressive force
f force de compression
d Druckkraft
e fuerza de compresión
n drukkracht

0263 compressive strength
f résistance à la compression
d Druckfestigkeit
e resistencia a compresión
n druksterkte

0264 compressive stress
f contrainte de compression
d Druckspannung
e tensión de compresión
n drukspanning

0265 compressive zone
f zone comprimée
d Druckzone
e zona de compresión
n drukzone

0266 compressor
f compresseur
d Kompressor
e compresor
n compressor

* **compute** *v* → 0170

0267 computer
f calculateur; calculatrice
d Rechner
e computador; calculadora electrónica
n computer

0268 concentrated load; point load
f charge concentrée; charge ponctuelle
d konzentrierte Belastung; Punktlast
e carga concentrado; carga punctual
n geconcentreerde belasting; puntlast

0269 concordant cable
f câble concordant
d konkordantes Spannglied
e cable concordante
n concordante kabel

0270 concrete
f béton
d Beton
e hormigón
n beton

0271 Concrete Association
f Association du béton
d Betonverein
e Asociación del Hormigón
n Betonvereniging

0272 concrete chute
f goulotte à béton
d Betonschüttrinne; Schüttrinne
e canaleta de hormigonado
n stortgoot

0273 concrete compacted by jolting
f béton compacté par choc; béton choqué
d Schockbeton
e hormigón picado; hormigón apisonado
n schokbeton

0274 concrete composition
f composition du béton
d Betonzusammensetzung
e composición del hormigón
n betonsamenstelling

0275 concrete cover
f recouvrement des armatures
d Betondeckung
e recubrimiento de hormigón
n betondekking

0276 concrete culvert
f ponceau en béton
d Betondurchlass
e atarjea de hormigón; alcantarilla de hormigón
n betonduiker

0277 concrete finisher
f finisseur de béton
d Betonfertiger
e acabadora
n betonafwerker

0278 concrete hinge
f articulation en béton
d Betongelenk
e rótula de hormigón; articulación de hormigón
n betonscharnier

0279 concrete mixer
f mélangeur à béton
d Betonmischer
e hormigonera
n betonmolen

0280 concrete mix(ture)
f mélange de béton

d Betonmischung
e mezcla de hormigón
n betonmengsel

0281 concrete on *v*
f bétonner sur
d betonieren auf
e hormigonar sobre
n aanstorten

0282 concrete pavement
f revêtement en béton
d Betondecke
e pavimento de hormigón
n betonverharding

0283 concrete placing; concreting; casting
f coulage du béton; bétonnage
d Betonieren; Betoneinbringung; Giessen
e colocación del hormigón; vertido del hormigón
n betonstorten

0284 concrete plant
f usine à béton
d Betonwerk
e central de hormigonado
n betoninstallatie

0285 concrete plywood (for formwork)
f contreplaqué de coffrage
d Sperrholz (for Betonschalung)
e encofrado de madera contrachapada
n betontriplex

0286 concrete probe (Humm's)
f sonde de Humm
d Betonsonde (nach Humm)
e sonda para el ensayo de consistencia del hormigón
n betonsonde (van Humm)

0287 concrete pump
f pompe à béton
d Betonpumpe
e bomba de hormigonado
n betonpomp

0288 concrete spacer block
f espaceur (en béton)
d Abstandhalter (aus Beton)
e separador
n betonblokje; betonnen afstandsblokje

0289 concrete technology
f technologie du béton
d Betontechnologie
e tecnología del hormigón
n betontechnologie

0290 concrete testing
f scléromètre
d Betonprüfhammer
e esclerómetro
n betonproefhamer

* **concreting** → 0283

0291 concreting hopper
f trémie de bétonnage
d Betonierbunker
e tolva de hormigonado
n storttrechter

0292 concreting skip
f benne de bétonnage
d Betonierkübel
e balde de hormigonado
n stortbak

0293 conditions of sale
f conditions de vente
d Lieferbedingungen
e condiciones de venta
n leveringsvoorwaarden

0294 cone
f cône
d Konus; Kegel
e cono
n conus

0295 conformity with the design dimensions
f conformité avec les cotes du projet
d Übereinstimmung mit den Abmessungen des Entwurfs
e conformidad con las dimensiones de proyecto
n maatvoering

0296 conical shell
f coque conique

d Kegelschale
e lámina cónica
n kegelschaal

0297 connecting nut
f écrou d'assemblage
d Verbindungsmutter
e tuerca de empalme
n koppelmoer

0298 connecting reinforcement
f armature de raccordement
d Anschlussbewehrung
e armadura de empalme; armadura pasante
n verbindingswapening

0299 conoidal shell
f coque conoïdale
d Konoidschale
e lámina conoidal
n konoïdeschaal

0300 consistency
f consistance
d Konsistenz
e consistencia
n consistentie

* **construction** → 0489

0301 construction joint
f joint de construction; joint de reprise
d Arbeitsfuge
e junta de construcción
n stortnaad

* **construction site** → 1083

0302 contact
f contact
d Kontakt
e contacto
n contact

0303 contact method
f méthode de contact
d Kontaktbauweise
e método de contacto
n contactmethode

0304 contact (sur)face
f face de contact; surface de contact
d Kontaktfläche
e superficie de contacto; cara de contacto
n contactvlak

0305 containing water
f contenant de l'eau
d wasserhaltig
e acuífero
n waterhoudend

0306 contaminate *v*; **pollute** *v*
f polluer
d verunreinigen
e contaminar
n verontreinigen

0307 contamination; pollution
f contamination; pollution
d Verunreinigung
e contaminación; polución
n verontreiniging

0308 content
f teneur
d Gehalt
e contenido
n gehalte

0309 contents
f contenu
d Inhalt
e índice; contenido
n inhoud

0310 continuity cable
f câble de continuité
d Kontinuitätskabel; Kabel zur Erzielung der Durchlaufwirkung
e cable de continuidad
n continuïteitskabel

0311 continuity moment
f moment de continuité
d Kontinuitätsmoment; Stützenmoment
e momento de continuidad
n overgangsmoment

0312 continuous
f continu
d durchlaufend; kontinuierlich
e contínuo
n doorgaand

0313 continuous beam
f poutre continue
d Durchlaufträger
e viga contínua
n doorgaande balk

0314 continuous grading
f granulométrie continue
d kontinuierliche Körnung
e granulometría contínua
n continue korrelverdeling

0315 continuous slab
f dalle continue
d durchlaufende Platte; Durchlaufplatte
e placa contínua
n doorgaande plaat

0316 continuous welding
f soudage en continu
d durchgehende Schweissung
e soldadura contínua
n continu lassen

0317 contract
f contrat
d Vertrag
e contrato; convenio
n contract

0318 contract price
f prix du marché
d Submissionssumme
e precio de contrato
n aannemingssom

0320 contracter
f entrepreneur
d Unternehmer
e contratista
n aannemer

0319 contracting firm
f entreprise
d Bauunternehmung
e contratista
n aannemersbedrijf

0321 contraction
f contraction
d Schrumpfung
e contracción; acortamiento
n contractie

0322 convex
f convexe
d konvex
e convexo
n convex

0323 cool *v*
f refroidir
d kühlen
e enfriar
n koelen

0324 cooperate *v*
f coopérer
d zusammenarbeiten
e cooperar
n samenwerken

0325 coordinate *v*
f coordonner
d koordinieren
e coordinar
n coördineren

0326 corbel
f console
d Konsole
e voladizo; ménsula; cartela
n console

0327 core
f noyau
d Kern
e núcleo
n kern

0328 core radius
f rayon du noyau
d Kernhalbmesser
e radio del núcleo
n kernstraal

0329 corner bar
f armature d'angle
d Eckstab
e barra de esquina
n hoekstaaf

0330 corrode *v*
f corroder
d korrodieren
e corroer
n corroderen

0331 corrosion
f corrosion
d Korrosion
e corrosión
n corrosie

0332 corrosion-resistant
f résistant à la corrosion
d korrosionsbeständig
e resistente a la corrosión
n corrosie-bestendig

0333 counter-current mixer
f malaxeur à contre-courant
d Gegenstrommischer
e amasadora de contra-corriente; hormigonera de contra-corriente
n tegenstroommolen

0334 couple
f couple
d Kräftepaar
e par
n koppel

0335 coupling
f couplage
d Kupplung
e empalme
n koppeling

0336 cover *v*
f couvrir
d bedecken
e recubrir; cubrir
n afdekken

* **cover plate** → 0159

0337 cover *v* **with sand**
f remblayer en sable
d mit Sand bedecken
e cubrir con arena
n afzanden

0338 crack
f fissure
d Riss
e fisura
n scheur

0339 crack pattern
f tracé des fissures
d Rissbild
e esquema de fisuración
n scheurpatroon

0340 crack spacing
f espacement des fissures
d Rissabstand
e separación entre fisuras
n scheurafstand

0341 crack width
f ouverture des fissures
d Rissbreite
e anchura de fisuras
n scheurwijdte

0342 cracking
f fissuration
d Reissen; Rissbildung
e fisuración
n scheurvorming

0343 cracking moment
f moment de fissuration
d Rissmoment
e momento de fisuración
n scheurmoment

0344 crane
f grue
d Kran
e grúa
n hijskraan

0345 creep
f fluage
d Kriechen
e fluencia
n kruip

0346 creep deformation
f déformation de fluage
d Kriechverformung
e deformación por fluencia
n kruipvervorming

0347 creep factor
f coefficient de fluage
d Kriechzahl
e factor de fluencia
n kruipfactor

0348 creep gradient
f gradient de fluage
d Kriechgefälle

e gradiente de fluencia
n kruipgradiënt

0349 creep modulus
f module de fluage
d Kriechmodul
e módulo de fluencia
n kruipmodulus

0350 crimp *v*
f onduler
d wellen
e doblar; ondular
n onduleren

0351 crimped wire
f fil ondulé
d gewellter Draht; Welldraht
e alambre ondulado
n geonduleerd draad

0352 criterion
f critère
d Kriterium
e criterio
n kriterium

0353 cross-beam
f poutre transversale
d Querbalken
e viga transversal; riostra
n dwarsbalk

0354 cross-girder
f entretoise
d Querträger
e riostra; viga de rigidez
n dwarsdrager

0355 cross-member
f traverse
d Riegel
e elemento transversal
n traverse

0356 cross-section
f section transversale
d Querschnitt
e sección transversal
n dwarsdoorsnede

0357 cross-wise reinforcement
f armature croisée
d kreuzweise Bewehrung
e armadura transversal
n kruisnet

0358 crushed aggregate
f agrégat concassé
d gebrochener Zuschlag
e árido machacado
n gebroken toeslagmateriaal

0359 crushed gravel
f gravier concassé
d Brechkies
e grava machacada
n gebroken grind

0360 crushed stone
f pierre concassée
d Schotter
e piedra machacada
n gebroken steen

0361 crushing (of aggregates)
f concassage (des agrégats)
d Brechen (von Zuschlagsmaterial)
e machaqueo (de áridos); trituración (de áridos)
n breken (van toeslagmaterialen)

0362 cube
f cube
d Würfel
e cubo
n kubus

0363 cube strength
f résistance sur cube
d Würfelfestigkeit
e resistencia en probeta cúbica
n kubussterkte

0364 curb; kerb
f bordure (de trottoir)
d Bordstein
e bordillo (de acera)
n trottoirband

0365 cure *v*
f curer
d nachbehandeln
e curar
n nabehandelen

0366 curing compound
f produit de cure

d Abdichtungsmittel
e compuesto de curado; producto de curado
n curing compound

0367 curvature
f courbure
d Krümmung
e curvatura
n kromming

0368 curvature gradient
f gradient de courbure
d Krümmungsverlauf
e gradiente de curvatura
n krommingsgradiënt

0369 curve
f courbe
d Kurve
e curva
n bocht

0370 curved
f courbé
d gebogen
e curvado
n gebogen

0371 curved cable
f câble courbé
d gekrümmtes Kabel
e cable curvado; cable de trazado curvo
n gebogen kabel

0372 cut *v*
f couper
d schneiden
e cortar
n knippen

0373 cutting schedule
f plan de découpage
d Schneidplan
e lista de corte de ferralla
n knipstaat

0374 cyclic loading
f charge cyclique
d periodische Belastung
e carga cíclica
n periodieke belasting

0375 cylinder
f cylindre
d Zylinder
e cilindro
n cilinder

0376 cylinder strength
f résistance sur cylindre
d Zylinderfestigkeit
e resistencia en probeta cilíndrica
n cilindersterkte

0377 cylindrical
f cylindrique
d zylindrisch
e cilíndrico
n cilindervormig

0378 cylindrical shell
f coque cylindrique
d Zylinderschale
e lámina cilíndrica
n cylindrische schaal

D

0379 dam
f barrage; grand barrage
d Talsperre
e presa; dique
n stuwdam

0380 damage *v*
f endommager
d beschädigen
e dañar; deteriorar
n beschadigen

0381 dead-end anchorage
f ancrage mort
d Blindverankerung
e anclaje pasivo; anclaje muerto
n blinde verankering

0382 dead load
f charge permanente
d Totlast
e carga permanente; carga muerta
n permanente belasting

0383 dead weight
f poids mort; poids propre
d Eigengewicht
e peso propio
n eigen gewicht

0384 deep beam
f poutre cloison
d wandartiger Träger
e viga de gran canto; viga pared
n hogewandligger

0385 deflect *v*
f dévier
d umlenken; ablenken
e desviar
n afbuigen

0386 deflect *v* **a prestressing wire**
f dévier un fil de précontrainte
d Spanndraht umlenken
e desviar un alambro de pretensado
n spandraad neerdrukken

0387 deflection
f flèche
d Durchbiegung
e flecha
n doorbuiging

0388 deform *v*
f déformer
d verformen
e deformar
n vervormen

0389 deformation
f déformation
d Verformung; Formänderung
e deformación
n vervorming; vormverandering

0390 deformed bar
f armature à adhérence renforcée
d Formstahl; profilierter Stahl
e barra corrugada
n geprofileerd staal

0391 de-icing salt
f sel antigel; sel de déverglaçage
d Tausalz; Auftausalz
e deshielo
n dooizout

0392 delivery
f livraison
d Übergabe
e suministro; entrega
n oplevering

0393 delivery pipe(line)
f conduite de refoulement
d Druckrohr(leitung)
e tubería de descarga
n persleiding

0394 density
f densité
d Dichte
e densidad
n dichtheid

* **depth** (of a structural section)
→ 0618

0395 design
f projet
d Entwurf
e proyecto; cálculo
n ontwerp

0396 design load(ing)
f charge de calcul
d Rechenlast
e carga de cálculo
n rekenbelasting

0397 design office
f bureau d'études
d Entwurfsbüro
e oficina de proyectos
n ontwerpbureau

0398 design value
f valeur de calcul
d Rechenwert
e valor de cálculo
n rekenwaarde

0399 designer
f projeteur
d Konstrukteur; Entwerfer
e proyectista
n constructeur; ontwerper

0400 destructive testing
f essai destructif
d zerstörende Prüfung
e ensayo destructivo
n destructief onderzoek

0401 detail
f détail
d Detail
e detalle
n detail

0402 diagonal
f diagonal
d diagonal
e diagonal
n diagonaal

0403 diagram; graph
f diagramme; graphique
d Diagramm; graphische Darstellung; Schaubild
e diagrama; gráfico
n diagram; grafiek

0404 diameter
f diamètre
d Durchmesser
e diámetro
n diameter

0405 diamond drill
f foreuse à diamant
d Diamantbohrer
e sonda de diamante
n diamantboor

0406 diaphragm
f diaphragme
d Querscheibe
e diafragma
n dwarsschot

0407 diaphragm pump
f pompe à membrane
d Membranpumpe
e bomba de membrana
n membraampomp

0408 diaphragm wall
f mur de fondation bétonné en tranchée
d Schlitzwand
e muro pantalla
n diepwand

0409 differential
f différentiel
d Differential
e diferencial
n differentiaal

0410 differentiate *v*
f différencier
d differenzieren
e diferenciar
n differentiëren

0411 dimension
f dimension; cote
d Mass; Dimension
e dimensión; medida
n afmeting; maat; dimensie

0412 dimensional deviation
f déviation dimensionnelle
d Massabweichung
e desviación dimensional
n maatafwijking

0413 directives
f directives
d Richtlinien
e directrices
n richtlijnen

0414 disc
f disque
d Scheibe
e disco
n schijf

0415 disconcordant cable
f câble non concordant
d nichtkonkordantes Spannglied
e cable no concordante
n disconcordante kabel

0416 discontinuity
f discontinuité
d Diskontinuität; Unstetigkeit
e discontinuidad
n discontinuïteit

0417 discontinuous grading; gap grading
f granulométrie discontinue
d diskontinuierliche Kornabstufung; Ausfallkörnung
e granulometría discontínua
n discontinue korrelverdeling

0418 dismantle *v*
f démonter
d demontieren; abmontieren
e desmontar; desmoldar; desarmar
n demonteren

0419 distance
f distance
d Abstand
e distancia
n afstand

0420 distemper *v*
f badigeonner
d tünchen
e destemplar
n sausen

0421 distribution reinforcement
f ferraillage de répartition
d Verteilungsbewehrung
e armadura de reparto
n verdeelwapening

0422 double-acting jack
f vérin à double effet
d doppelwirkende Presse
e gato de doble acción; gato de doble efecto
n dubbelwerkende vijzel

0423 double bend
f courbure double
d Doppelkrümmung
e doble curvatura
n dubbele bocht

0424 dowel
f goujon
d Dübel
e clavija; pasador
n deuvel

* **draftsman** → 0426

0425 dragline
f dragline
d Schürfkübelbagger; Schleppschaufelbagger
e dragalina
n sleepschop; dragline

0426 draughtsman; draftsman
f dessinateur
d Zeichner
e delineante
n tekenaar

0427 draw-bench
f banc d'étirage
d Ziehbank
e banco de enderezado
n trekbank

0428 drawing
f dessin
d Zeichnung; Plan
e plano; dibujo; diseño
n tekening

0429 drill core
f carotte
d Bohrkern
e broca de perforación; barrena de perforación
n boorkern

0430 driving stress
f sollicitation au battage (pieu)
d Rammbeanspruchung
e tensión de hinca
n heispanning

* **drop** → 0804

0431 drop-in girder; suspended beam
f poutre suspendue
d Einhängträger
e viga colgante
n inhangligger

0432 dry *v*
f sécher
d austrocknen
e desecar
n uitdrogen

0433 dry concrete
f béton sec
d erdfeuchter Beton
e hormigón seco
n aardvochtig beton

0434 dry dock
f cale sèche; forme de radoub
d Trockendock
e dique seco
n droogdok

0435 drying shrinkage
f retrait au séchage
d Austrocknungsschwinden
e retracción de secado
n uitdrogingskrimp

0436 ductube
f ductube; tuyau en caoutchouc gonflable
d aufblasbarer Gummischlauch für Kanäle
e ductube
n ductube

0437 dummy joint
f joint aveugle
d Scheinfuge
e simulado
n schijnvoeg

0438 dump body
f benne basculante
d Kippmulde
e cuerpo basculante
n kipbak

0439 dune sand
f sable de dune
d Dünensand
e arena de duna
n duinzand

0440 durable
f durable
d beständig; dauerhaft
e duradero; durable
n duurzaam

0441 dynamic load(ing)
f charge dynamique
d dynamische Belastung
e carga dinámica
n dynamische belasting

0442 dynamic testing
f essai dynamique
d dynamische Prüfung
e ensayo dinámico
n dynamisch onderzoek

0443 dynometer
f dynamomètre
d Dynamometer
e dinamómetro
n dynamometer

E

0444 earth pressure
f poussée des terres
d Erddruck
e empuje de tierras
n gronddruk

0445 earth slope
f talus
d Erdböschung
e pendiente; talud
n talud

0446 earthquake load
f charge sismique
d Erdbebenbelastung
e carga sísmica
n aardbevingsbelasting

0447 eccentric load(ing)
f charge excentrée
d exzentrische Belastung; ausmittige Belastung
e carga excéntrica
n excentrische belasting

0448 eccentricity
f excentricité
d Exzentrizität; Ausmittigkeit
e excentricidad
n excentriciteit

0449 edge
f bord
d Rand
e borde
n rand

0450 edge beam
f poutre de rive
d Randträger
e viga de borde
n randbalk

0451 edge moment
f moment au bord
d Randmoment
e momento de borde
n randmoment

0452 effective length (in buckling)
f longueur de flambage
d Knicklänge
e longitud efectiva (en pandeo)
n kniklengte

0453 effective width
f largeur effective
d mitwirkende Breite
e anchura eficaz
n medewerkende breedte

0454 efficiency
f rendement
d Wirkungsgrad
e eficiencia; rendimiento
n rendement

0455 efflorescence
f efflorescence
d Ausblühung
e eflorescencia
n kalkuitslag

0456 elastic
f élastique
d elastisch
e elástico
n elastisch

0457 elastic deformation
f déformation élastique
d elastische Verformung
e deformación elástica
n elastische vervorming

0458 elastic limit
f limite élastique
d Elastizitätsgrenze
e límite elástico
n elasticiteitsgrens

0459 elastically supported girder
f poutre sur appuis élastiques
d elastisch gestützter Träger
e viga elásticamente apoyada
n elastisch ondersteunde ligger

0460 elasticity
f élasticité
d Elastizität
e elasticidad
n elasticiteit

0461 electric welding
f soudure électrique
d elektrisches Schweissen

e soldadura eléctrica
n electrisch lassen

0462 electrothermal hardening
f durcissement par chauffage électrique
d elektrothermische Härtung
e endurecimiento electrotérmico
n electrothermisch verharden

0463 electrothermal prestressing
f précontrainte par chauffage électrique
d elektrothermische Vorspannung
e pretensado electrotérmico
n electrothermische voorspanning

0464 elevation
f élévation
d Ansicht; Aufriss
e alzado
n aanzicht

0465 ellipse of inertia
f ellipse d'inertie
d Trägheitsellipse
e elipse de inercia
n traagheidsellips

0466 elongate *v*
f étirer; allonger
d dehnen; strecken
e alargar; prolongar
n rekken; verlengen

0467 elongation
f allongement
d Dehnung
e alargamiento; extensión
n rek

• **elongation at fracture** → 1314

0468 empirical
f empirique
d empirisch
e empírico
n empirisch

0469 encase *v*
f enrober
d umhüllen
e cubrir; forrar; encajar
n omhullen

0470 encasement
f enrobage
d Umhüllung
e conducto
n omhulling

0471 end bearing
f appui extrême
d Endauflager(ung)
e apoyo extremo
n eindoplegging

0472 end block
f bloc d'about
d Endblock
e bloque extremo; cabeza de anclaje
n eindblok; ankerblok

0473 end-block reinforcement
f ferraillage d'about
d Endblockbewehrung
e armadura de la cabeza de anclaje
n kopnet

0474 engineer
f ingénieur
d Ingenieur
e ingeniero
n ingenieur

0475 envelop *v*
f envelopper
d ummanteln
e envolver; enfundar; envainar
n omhullen

0476 envelopment
f enveloppement
d Ummantelung
e envoltura; funda; recubrimiento
n omhulling

0477 equilibrium
f équilibre
d Gleichgewicht
e equilibrio
n evenwicht

0478 equilibrium condition
f condition d'équilibre
d Gleichgewichtsbedingung
e condición de equilibrio
n evenwichtsvoorwaarde

0479 equipment
f matériel
d Geräte
e equipo; utillaje; herramientas
n materiëel

* **erect framework** *v* → 1076

0480 erection
f montage
d Montage
e montaje; erección
n montage

0481 erection load(ing)
f charge de montage
d Montagebelastung
e carga de montage
n montagebelasting

0482 estimate
f devis estimatif; évaluation
d Kosten(vor)anschlag
e estimación; valoración
n begroting

0483 etch *v*
f corroder; attaquer à l'acide
d ätzen
e decapar; desoxidar
n etsen

0484 etching batch
f bain corrodant
d Ätzbad
e baño de decapado
n etsbad

0485 etching test
f essai par corrosion
d Ätzversuch
e ensayo de decapado
n etsproef

0486 examination
f examen
d Untersuchung; Prüfung
e examen; investigación; inspección
n onderzoek

0487 excavate *v*
f excaver
d ausheben; baggern
e excavar
n ontgraven

0488 excavator
f excavateur
d Trockenbagger
e excavadora
n excavator; graafmachine

0489 execution; construction
f exécution; réalisation
d Ausführung; Bauausführung
e ejecución; construcción; estructura
n uitvoering; bouw

0490 expanded metal
f métal déployé
d Streckmetall
e metal desplegado
n gerekt metaal

0491 expansion
f dilatation; expansion
d Ausdehnung; Dehnung
e expansión; dilatación
n dilatatie; uitzetting (fysisch)

0492 expansion joint
f joint de dilatation
d Dehnfuge; Raumfuge
e junta de dilatación
n dilatatievoeg; uitzettingsvoeg

0493 expansive cement
f ciment expansif
d Quellzement
e cemento expansivo
n zwelcement

0494 experiment
f expérience; essai
d Versuch
e experimento; ensayo
n proef; experiment

0495 explosion load(ing)
f charge d'explosion
d Explosionsbelastung
e carga originada por explosión
n explosiebelasting

0496 extend *v*
f allonger
d verlängern
e alargar; prolongar
n verlengen

0497 extensometer
f extensomètre
d Extensometer
e medidor de deformaciones unitarias; comparador
n extensometer

0498 external; outside
f externe; extérieur
d äusserlich; auswendig; aussen-
e externo; exterior
n uitwendig

0499 external cable
f câble extérieur
d Aussenkabel
e cable exterior
n uitwendige kabel

0500 external prestress
f précontrainte externe
d Aussenvorspannung
e pretensado externo
n uitwendige voorspanning

0501 external span
f travée externe
d Aussenfeld
e tramo exterior
n buitenveld

0502 external vibrator
f vibrateur externe
d Aussenrüttler
e vibrador externo
n bekistingsvibrator

0503 extrapolate *v*
f extrapoler
d extrapolieren
e extrapolar
n extrapoleren

0504 extreme fibre
f fibre extrême
d Randfaser
e fibra extrema
n uiterste vezel

0505 extreme (fibre) stress
f contrainte dans la fibre extrême
d Randspannung
e tensión de borde
n randspanning

0506 extruded concrete
f béton extrudé
d Strangpressbeton
e hormigón extruído
n strengpersbeton; geëxtrudeerd beton

F

0507 fabric mat (spot-welded)
f treillis soudé
d Baustahlmatte; Bewehrungsmatte (punktgeschweisst)
e malla de armadura (electrosoldada)
n wapeningsnet (gepuntlast)

0508 failure
f rupture
d Bruch
e rotura; agotamiento
n breuk

0509 failure hypothesis
f hypothèse de rupture
d Bruchhypothese
e hipótesis de rotura
n breukhypothese

0510 failure limit
f limite de rupture
d Bruchgrenze
e límite de rotura
n breukgrens

0511 failure load
f charge de rupture
d Bruchlast
e carga de rotura
n bezwijkbelasting; breukbelastung

0512 failure mechanism
f mécanisme de rupture
d Bruchmechanismus
e mecanismo de rotura
n breukmechanisme

0513 failure moment
f moment de rupture
d Bruchmoment
e momento de rotura
n breukmoment

0514 failure stage
f état de rupture
d Bruchstadium
e estado de rotura
n breukstadium

0515 falsework
f étaiement (de coffrage)
d Schalgerüst
e andamio
n steiger (als bekisting); bekistingssteiger

0516 fan type anchorage
f ancrage en éventail
d Fächerverankerung
e anclaje en abanico
n spreidverankering

0517 fatigue
f fatigue
d Ermüdung
e fatiga
n vermoeiing

0518 fatigue strength
f résistance à la fatigue
d Ermüdungsfestigkeit
e resistencia a la fatiga
n vermoeiingssterkte

0519 ferrite
f ferrite
d Ferrit
e ferrita
n ferriet

0520 fibre
f fibre
d Faser
e fibra
n vezel

0521 fictitious
f fictif
d fiktiv
e ficticio
n fictief

0522 fillet
f baguette d'angle
d Eckleiste
e filete; chaflán; bisel
n hoeklat

0523 fine aggregate
f agrégat fin
d Feinzuschlag
e árido fino
n fijn toeslagmateriaal

0524 fine gravel
f gravier fin
d Feinkies
e grava fina; gravilla
n fijn grind

0525 fineness modulus
f module de finesse
d Feinheitsmodul
e módulo de finura
n fijnheidmodulus

0526 fineness of grinding
f finesse de broyage
d Mahlfeinheit
e finura de molido
n maalfijnte

0527 finish *v*
f finir
d fertigbearbeiten; fertigen
e acabar
n afwerken

0528 finish by screeding board
f finissage à la latte de réglage
d Fertigen mit der Abziehlatte
e acabado especial de la superficie del hormigón
n afwerken onder de rij

0529 fire duration
f durée du feu
d Branddauer
e duración del incendio
n brandduur

0530 fire-resistant
f résistant au feu
d feuerbeständig
e resistente al fuego
n brandbestendig

0531 fire safety
f sécurité au feu
d Feuersicherheit
e seguridad al fuego
n brandveiligheid

0532 fix *v*
f encastrer
d einspannen
e atornillar; apretar; consolidar
n inklemmen

0533 fixed-end moment
f moment d'encastrement parfait
d Volleinspannmoment
e momento de empotramiento
n primair moment

0534 fixing moment
f moment d'encastrement
d Einspannmoment
e momento de empotramiento
n inklemmingsmoment

0535 flange (of a beam)
f membrure (d'une poutre); semelle
d Flansch; Gurt
e ala (de una viga)
n flens (van een balk)

0536 flange width
f largeur de membrure
d Gurtbreite
e anchura del ala
n flensbreedte

0537 flanged nut
f écrou à collet
d Bundmutter
e tuerca de mariposa
n kraagmoer

0538 flat
f plat
d flach
e plano; liso
n vlak; plat

0539 flexural strength
f résistance à la flexion
d Biegezugfestigkeit
e resistencia a flexión
n buigtreksterkte

0540 floating dock
f dock flottant
d Schwimmdock
e dique flotante
n drijvend dok

0541 floor loading
f charge de plancher
d Deckenbelastung
e carga de piso
n vloerbelasting

0542 flush *v* (with water)
f laver (à l'eau)
d durchspülen (mit Wasser)
e limpiar con chorro de agua
n doorspoelen

0543 flyover
f passage supérieur
d Kreuzungsbauwerk; Überführungsbauwerk
e viaducto; paso superior
n viaduct

0544 foamed concrete
f béton-mousse
d Schaumbeton
e hormigón espumado
n schuimbeton

* **foil** (metal) → 1058

0545 folding wedges
f coins mâtés
d Doppelkeile
e cuña doble
n dubbele wig

0546 force
f force
d Kraft
e fuerza
n kracht

0547 formula
f formula
d Formel
e fórmula
n formule

0548 formwork; shuttering
f coffrage
d Schalung
e encofrado; molde
n bekisting

0549 formwork oil; mould oil
f huile de coffrage
d Schalungsöl
e desencofrante
n bekistingsolie

0550 formwork vibrator
f vibrateur de coffrage
d Schalungsrüttler
e vibrador de encofrado; vibrador de molde
n bekistingsvibrator

0551 foundation
f foundation
d Fundierung; Fundament; Gründung
e cimentación
n fundering

0552 foundation slab
f dalle de fondation
d Fundierungsplatte
e placa de cimentación
n funderingsplaat

0553 fracture load
f charge de rupture
d Bruchlast
e carga de rotura
n breukbelasting

0554 fractured
f rompu
d gebrochen
e roto; fracturado
n gebroken

0555 framework
f ossature
d Skelett
e esqueleto; armazón
n skelet

0556 frequency
f fréquence
d Frequenz; Häufigkeit
e frecuencia
n frequentie

0557 fresh concrete; freshly mixed concrete
f béton frais
d Frischbeton
e hormigón fresco
n betonspecie

* **freshly mixed concrete** → 0557

0558 frost-resistant
f résistant au gel
d frostbeständig
e resistente a la helada; resistente

al hielo
n vorstbestendig

0559 funicular polygon
f polygone funiculaire
d Seilzug
e polígono funicular
n stangenveelhoek

0560 furnace clinker concrete
f béton de mâchefer
d Schlackenbeton
e hormigón de escoria
n slakkenbeton

G

0561 galvanize *v*
f galvaniser
d verzinken
e galvanizar
n verzinken

* **gap grading** → 0417

0562 gas welding
f soudage au gaz
d Gasschweissung
e soldadura autógena
n autogeen lassen

0563 gasket; seal
f garniture
d Dichtung
e empaquetadura; selladura
n pakking; dichting

0564 geometric(al)
f géométrique
d geometrisch
e geométrico
n geometrisch

0565 girder
f poutre
d Träger; Balken
e viga
n ligger; bint

0566 girder bridge
f pont à poutres
d Balkenbrücke
e puente de vigas
n balkbrug

0567 glue
f colle
d Leim
e cola de pegar
n lijm

0568 glueing method
f méthode par collage
d Klebeverfahren
e método de encolado
n lijmmethode

0569 Goliath crane
f grue portique
d Brückenkran; Portalkran
e pórtico de apoyo; pórtico de carga
n portaalkraan; traverse

0570 grade *v*
f calibrer
d abstufen nach Korngrösse; sieben
e gradar
n graderen

0571 grade; quality class
f classe de qualité
d Güteklasse
e grado
n kwaliteitsklasse

0572 grading
f granulométrie
d Korngrössenverteilung
e granulometría
n korrelverdeling

0573 grading curve
f courbe granulométrique
d Sieblinie
e curva granulométrica
n zeefkromme

0574 granite chippings
f concassé de granit
d Granatsplitt
e gravilla de granito
n granietsplit

0575 granulate *v*
f granuler
d granulieren
e granular
n granuleren

0576 granulometry
f granulométrie
d Kornaufbau
e granulometría
n granulometrie

* **graph** → 0403

0577 graphite
f graphite
d Graphit
e grafito
n grafiet

0578 graphostatics
f statique graphique
d graphische Statik
e grafostático
n grafostatica

0579 gravel
f gravier
d Kies
e grava
n grind

0580 gravel concrete
f béton de gravier
d Kiesbeton
e hormigón de grava
n grindbeton

0581 gravel pocket
f nid de cailloux
d Kiesnest
e coquera
n grindnest

0582 gravity
f gravité
d Schwerkraft
e gravedad
n zwaartekracht

0583 grid
f gril(le)
d Netz
e rejilla; emparrillado
n stramien

0584 grillage floor
f plancher à poutrage en gril
d Trägerrostdecke
e emparrillado de piso
n roostervloer

0585 grip anchorage
f ancrage par serrage
d Klemmverankerung
e grapa de anclaje
n klemverankering

0586 ground surface
f surface du terrain
d Geländeoberfläche
e superficie del terreno
n maaiveld; terreinoppervlakte

0587 ground water
f eau phréatique
d Grundwasser
e agua subterránea
n grondwater

0588 group of bars
f groupe de barres
d Stabbündel
e grupo de barras
n staafgroep

0589 grout
f coulis; mortier d'injection
d Einpressmörtel; Auspressmörtel
e lechada
n injectiespecie

0590 grout mixer
f mélangeuse à mortier
d Mörtelmischer; Zementmörtelmischer
e mezcladora del producto de inyección
n mengketel (voor injectiespecie)

0591 grouting
f injection
d Einpressen; Verpressen; Injektion
e inyección
n injectie

0592 grouting cup
f cloche d'injection
d Injizierglocke
e embudo de inyección
n injectieklok

0593 grouting hole
f orifice d'injection
d Einpressöffnung; Injizieröffnung
e orificio de inyección
n injectie-opening

0594 grouting hose
f tuyau d'injection
d Injektionsschlauch; Einpresschlauch
e tubo de inyección; manguera de inyección
n injectieslang

0595 grouting machine
f machine d'injection

d Injektionsgerät; Einpressgerät
e máquina de inyección
n injectieapparaat

0596 grouting pump
f pompe d'injection
d Injektionspumpe; Einpresspumpe
e bomba de inyección
n injectiepomp

0597 grouting rod
f lance d'injection
d Injektionslanze; Einpresslanze
e barra de inyección
n injectielans

0598 gypsum
f gypse
d Gips
e yeso
n gips

H

0599 hair crack
f fissure capillaire
d Haarriss
e fisura capilar
n haarscheur

0600 hair felt
f feutre poilu
d Haarfilz
e fieltro de pelo
n haarvilt

0601 hairpin
f épingle à cheveux
d Haarnadel
e horquilla
n haarspeld

0602 handling reinforcement
f armature de montage
d Montagebewehrung
e armadura de montaje
n montagebewapening

0603 harden *v*
f durcir
d erhärten
e endurecer
n verharden

0604 harden *v* (steel)
f tremper (acier)
d härten (Stahl)
e endurecer (acero)
n harden (staal)

0605 hardened cement paste
f pâte de ciment durcie
d Zementstein
e roca cementada
n cementsteen

0606 hardening
f durcissement
d Erhärten
e endurecimiento
n verharding

0607 hardening test
f essai de durcissement
d Erhärtungsprüfung
e ensayo de endurecimiento
n verhardingsproef

0608 hardness
f dureté
d Härte
e dureza
n hardheid

0609 haunch (on beam)
f gousset (d'une poutre)
d Voute (eines Balkens)
e cartabón; cartela
n console; balkverhoging

0610 haunching concrete (around a pipe)
f béton d'enrobage (d'un tuyau)
d Betonummantelung (eines Rohres)
e hormigón para recibido (alrededor de una tubería)
n vulbeton (om een buisleiding)

0611 headroom
f hauteur libre
d lichte Höhe
e altura libre
n vrije hoogte

0612 heat
f chaleur
d Hitze
e calor
n warmte

0613 heat insulation
f isolation thermique
d Wärmeisolierung
e aislamiento térmico
n warmte-isolatie

0614 heat of hydration
f chaleur d'hydratation
d Hydratationswärme
e calor de hidratación
n hydratatiewarmte

0615 heat of setting
f chaleur de prise
d Abbindewärme
e calor de fraguado
n bindingswarmte

0616 heat-treated steel
f acier traité thermiquement
d wärmebehandelter Stahl; vergüteter Stahl

e acero sometido a tratamiento térmico
n nabehandeld staal; veredeld staal

0617 heavy concrete
f béton lourd
d Schwerbeton
e hormigón pesado
n zwaartebeton

0618 height (general); **depth** (of a structural section)
f hauteur (d'une section)
d Höhe (eines Schnittes)
e altura (de una sección)
n hoogte (van een doorsnede)

* **high-alumina cement** → 0031

0619 high-pressure hose
f flexible à haute pression
d Hochdruckschlauch
e tubería de alta presión; conduccíon forzada
n hogedrukslang

0620 high-rise building
f immeuble élevé
d Hochhaus
e edificio alto
n hoogbouw

0621 high-strength concrete
f béton à haute résistance
d hochfester Beton
e hormigón de alta resistencia
n hoogwaardig beton

0622 high-tensile steel
f acier à haute résistance
d hochfester Stahl
e acero de alta resistencia
n hoogwaardig staal

0623 hinge
f articulation; rotule
d Gelenk
e articulación; rótula
n scharnier

0624 hinged bearing
f appui articulé
d Gelenklager
e apoyo articulado
n scharnieroplegging

0625 hogging (of beam)
f cambrure (d'une poutre)
d Aufwölbung (eines Trägers)
e contraflecha (de viga)
n opbuigen (van een balk)

0626 honeycomb structure
f structure alvéolaire
d Wabenstruktur
e estructura en colmena
n raatwerk

0627 hook
f crochet
d Haken
e gancho
n haak

0628 hoop reinforcement
f frettage
d Umschnürungsbewehrung
e zuncho; armadura helicoidal
n omwikkelingswapening

0629 horizontal
f horizontal
d horizontal; waagerecht
e horizontal
n horizontaal; waterpas

0630 horizontal section
f section horizontale
d Horizontalschnitt
e sección horizontal
n horizontale doorsnede

0631 horsepower; HP
f cheval-vapeur; CV
d Pferdestärke; PS
e caballo vapor; CV
n paardekracht; PK

0632 hot(-dip) galvanizing
f galvanisation à chaud
d Feuerverzinken
e galvanización por immersión en caliente
n thermisch verzinken

0633 hot-rolled steel
f acier laminé à chaud
d warmgewalzter Stahl
e acero laminado en caliente
n warmgewalst staal

0634 hot-rolled wire
f fil laminé à chaud
d warmgewalzter Draht
e alambre laminado en caliente
n warmgewalst draad

* **HP** → 0631

0635 humidity
f humidité
d Feuchtigkeit
e humedad
n vochtigheid

0636 hydration
f hydratation
d Hydratation
e hidratación
n hydratatie

0637 hydration shrinkage
f contraction par hydratation
d Schrumpfung infolge Hydratation
e retracción de hidratación
n hydrateringskrimp

0638 hydraulic binding agent
f liant hydraulique
d hydraulisches Bindemittel
e conglomerante hidráulico; agente conglomerante hidráulico
n hydraulisch bindmiddel

0639 hydrogen embrittlement
f fragilisation par l'hydrogène
d Wasserstoffversprödung
e fragilización por hidrógeno
n waterstofbroosheid

0640 hygroscopic
f hygroscopique
d hygroskopisch
e higroscópico
n hygroscopisch

0641 hyperbolic shell
f coque hyperbolique
d hyperbolische Schale
e lámina hiperbólica
n hypparschaal

* **hyperstatic** → 1157

0642 hypothesis
f hypothèse
d Hypothese
e hipótesis
n hypothese

I

0643 ice-loading
f charge de glace
d Eisbelastung
e carga de hielo
n ijsbelasting

0644 immediate deformation; instantaneous deformation
f déformation instantanée
d Sofortverformung
e deformación instantánea
n onmiddellijke vervorming

0645 immediate strain; instantaneous strain
f déformation instantanée
d Sofortdehnung
e deformación unitaria instantánea
n onmiddellijke rek (specifieke vormverandering)

0646 impact
f effort de choc
d Stoss(beanspruchung)
e impacto; choque
n stoot

0647 impact coefficient
f coefficient de choc
d Stossbeiwert
e coeficiente de impacto
n stootcoëfficient

0648 impact load(ing)
f effort de choc
d Stossbelastung
e carga de ımpacto
n stootbelasting

0649 impact strength
f résistance au choc
d Stossfestigkeit
e resistencia al impacto
n stootvastheid

* **impact value** → 0814

0650 impermeable
f imperméable
d undurchlässig
e impermeable
n ondoorlatend

* **imposed load(ing)** → 0723

0651 inclination
f inclinaison
d Neigung
e inclinación; buzamiento
n helling (hoek)

0652 indented wire
f fil à empreintes
d profilierter Draht
e alambre grafilado
n gedeukt draad

0653 inertia
f inertie
d Trägheit
e inercia
n traagheid

0654 initial
f initial
d anfänglich
e inicial
n initiëel

0655 initial strength
f résistance initiale
d Anfangsfestigkeit
e resistencia inicial
n aanvangssterkte

0656 initial stress
f contrainte initiale
d Anfangsspannung
e tensión inicial
n aanvangsspanning

0657 inject *v*
f injecter
d injizieren; einpressen
e inyectar
n injecteren

0658 injection
f injection
d Injektion
e inyección
n injectie

0659 inner face (of wall)
f face interne (d'un mur)
d Innenfläche (einer Wand)
e cara interna (de un muro)

n binnenoppervlakte (van een wand)

0660 inner span; internal span
f travée intermédiaire
d Innenfeld
e tramo interno; tramo interior
n binnenveld

0661 insert *v*
f introduire; enfoncer
d einbauen; einführen
e introducir; hincar
n inbrengen; insteken

* **inside** → 0669

0662 in-situ concrete
f béton coulé en place
d Ortbeton
e hormigón vertido in situ
n ter plaatse gestort beton

0663 inspect *v*
f inspecter
d prüfen; kontrollieren
e inspeccionar; examinar
n keuren

0664 inspector
f inspecteur
d Aufseher
e inspector
n opzichter

0665 instability
f instabilité
d Instabilität
e inestabilidad
n instabiliteit

* **instantaneous deformation** → 0644

* **instantaneous strain** → 0645

0666 integrate *v*
f intégrer
d integrieren
e integrar
n integreren

0667 interaction diagram
f diagramme d'interaction
d Zwischenwirkungsdiagramm
e diagrama de interacción
n interactie-diagram

0668 intermediate support
f support intermédiaire
d Zwischenstütze
e apoyo intermedio
n tussensteunpunt

0669 internal; inside
f interne; à l'intérieur
d innen; inwendig
e interior; dentro; interno
n inwendig

0670 internal diameter
f diamètre interne
d Innendurchmesser
e diámetro interior
n inwendige diameter

0671 internal dimension
f dimension interne
d Innenmass
e dimensión interior
n inwendige afmeting

0672 internal friction
f frottement interne
d innere Reibung
e rozamiento interno
n inwendige wrijving

* **internal span** → 0660

0673 internal vibrator
f pervibrateur; vibrateur interne
d Innenrüttler
e vibrador interno; vibrador de aguja
n trilnaald

0674 interpolate *v*
f interpoler
d interpolieren
e interpolar
n interpoleren

0675 interspace
f espace intermédiaire
d Zwischenraum
e espacio; intervalo
n tussenruimte

0676 investigation
f investigation
d Untersuchung
e investigación
n onderzoek

0677 invitation to tender
f appel d'offres
d Ausschreibung
e anuncio de subasta
n aanbesteding

0678 iron
f fer
d Eisen
e hierro
n ijzer

0679 iron wire
f fil de fer
d Eisendraht
e alambre de hierro
n ijzerdraad

0680 irreversible
f irréversible
d nicht umkehrbar
e irreversible
n irreversibel

0681 isotropic
f isotrope
d isotrop
e isotrópico; isótropo
n isotroop

0682 iterate *v*
f itérer
d iterieren
e iterar; repetir
n itereren

J

0683 jack
f vérin
d Presse; Spannpresse
e gato
n vijzel

* **job site** → 1083

0684 join *v*
f assembler
d zusammenfügen; fügen
e unir; enlazar; conectar
n verbinden; samenvoegen

0685 joint
f joint
d Fuge
e junta
n voeg

0686 jolting table
f table à secousses
d Schocktisch
e mesa de sacudidas
n schoktafel

K

0687 keep wet *v*
f maintenir humide
d nass halten; feucht halten
e mantener húmedo
n nat houden

* **kerb** → 0364

0688 kink (in a cable)
f boucle (dans un câble)
d Klanke (in einem Kabel)
e coca (en un cable); enredo (en un cable)
n knik (in kabel)

L

0689 laboratory
f laboratoire
d Laboratorium; Labor
e laboratorio
n laboratorium

* **labourer** → 1408

0690 landing
f palier
d Treppenabsatz; Podest
e rellano; descansillo; meseta
n bordes; overloop

0691 lap joint
f joint par recouvrement
d Überlappungsstoss
e empalme por solapo
n overlappingskracht

0692 lap length
f longueur de recouvrement
d Überlappungslänge
e longitud del empalme por solapo
n overlappingslengte

0693 lap weld
f soudure à recouvrement
d Überlappnaht
e soldadura por solapo
n overlappingslas

0694 lateral area
f surface latérale
d Mantelfläche
e área lateral
n manteloppervlak

0695 lateral contraction; transverse contraction
f contraction transversale
d Querkontraktion
e contracción lateral
n dwarscontractie

0696 lattice framework; truss
f treillis; construction en treillis
d Fachwerk
e celosía; enrejado
n vakwerk

0697 lattice girder; truss
f poutre à treillis
d Fachwerkträger; Gitterträger
e viga en celosía; viga triangulada
n vakwerkligger

0698 lattice theory
f théorie du treillis
d Fachwerktheorie
e teoría de las bielas
n vakwerktheorie

0699 leach (out) *v*
f lessiver
d auslaugen
e lixiviar
n uitlogen

0700 lead
f plomb
d Blei
e plomo
n lood

0701 lean concrete
f béton maigre
d Magerbeton
e hormigón pobre
n schraal beton

0702 length
f longueur
d Länge
e longitud; largo
n lengte

* **letting** (of a contract) → 0069

0703 level *v*
f régler; égaliser
d abgleichen; abziehen
e nivelar; enrasar
n afrijen; waterpas egaliseren; horizontaal egaliseren

0704 lever arm
f bras de levier
d Hebelarm
e brazo de palanca
n hefboomsarm

0705 lifting appliance
f appareil de levage
d Hebezeug

e aparato de elevación
n hijswerktuig

0706 lifting eye
f œil de suspension
d Aufhängeauge
e gancho de elevación
n hijsoog

0707 lightweight aggregate
f agrégat léger
d Leichtzuschlag
e árido ligero
n licht toeslagmateriaal

0708 lightweight concrete
f béton léger
d Leichtbeton
e hormigón ligero
n lichtbeton

0709 lime
f chaux
d Kalk
e cal
n kalk

0710 lime efflorescence
f efflorescence de chaux
d Kalkausblühung
e eflorescencia caliza
n kalkuitslag

0711 limestone
f calcaire
d Kalkstein
e caliza; piedra caliza
n kalksteen

0712 limit of proportionality
f limite de proportionnalité
d Proportionalitätsgrenze
e límite de proporcionalidad
n evenredigheidsgrens; proportionaliteitsgrens

0713 limit state
f état-limite
d Grenzzustand
e estado límite
n grenstoestand

0714 limit state of crack width
f état-limite de l'ouverture des fissures
d Grenzzustand der Rissbreite
e estado límite de fisuración controlada
n grenstoestand van scheurwijdte

0715 limit state of cracking
f état-limite de fissuration
d Grenzzustand der Rissbildung
e estado límite de fisuración
n grenstoestand van scheurvorming

0716 limit state of decompression
f état-limite de décompression
d Grenzzustand der Entlastung
e estado límite de descompresión
n grenstoestand van ontspanning

0717 line of action
f ligne d'action
d Wirkungslinie
e línea de acción
n werklijn

0718 line of thrust
f ligne de poussée
d Drucklinie
e línea de empuje
n druklijn

0719 linear
f linéaire
d linear; axial
e lineal
n lineair

0720 linear bearing
f appui linéaire
d Linienlager
e apoyo lineal
n lijnoplegging

0721 linear load
f charge linéaire
d Linienlast
e carga lineal
n lijnlast

0722 lintel
f linteau
d Sturz
e dintel
n latei

0723 live load(ing); imposed load(ing); superimposed load(ing)
f charge utile
d Nutzlast; Verkehrslast
e carga útil; sobrecarga
n nuttige belasting

0724 load
f charge
d Last
e carga
n last

0725 load-bearing wall
f mur porteur
d Tragwand
e muro de carga
n dragende muur

0726 load(-carrying) capacity
f capacité portante
d Tragvermögen
e capacidad de carga; capacidad portante
n draagvermogen

0727 load cell
f cellule manométrique
d Druckdose; Lastmessdose
e célula de carga
n drukdoos

0728 load combination
f combinaison de charges
d Belastungskombination
e combinación de cargas
n belastingscombinatie

0729 load train
f train de charge
d Lastenzug
e sistema de carga
n laststelsel

0730 loading
f chargement
d Belastung
e carga
n belasting

0731 lock *v*
f bloquer
d blockieren
e bloquear
n blokkeren

0732 lock-nut
f contre-écrou
d Gegenmutter
e contratuerca
n borgmoer

0733 locking jack
f vérin de bloquage
d Blockierpresse
e gato con dispositivo de bloqueo
n blokkeervijzel

0734 locking ram
f piston de bloquage
d Blockierkolben
e émbolo con dispositivo de bloqueo
n blokkeerplunjer

0735 logarithmic
f logarithmique
d logarithmisch
e logarítmico
n logarithmisch

0736 longitudinal bar
f barre longitudinale
d Längsstab
e barra longitudinal
n langsstaaf

0737 longitudinal crack
f fissure longitudinale
d Längsriss
e fisura longitudinal
n langsscheur

0738 longitudinal force
f force longitudinale
d Längskraft
e fuerza longitudinal
n langskracht

0739 longitudinal girder
f poutre longitudinale
d Längsträger
e viga longitudinal
n langsligger

0740 longitudinal prestress(ing)
f précontrainte longitudinale
d Längsvorspannung
e pretensado longitudinal
n langsvoorspanning

0741 longitudinal reinforcement
f armature longitudinale
d Längsbewehrung
e armadura longitudinal
n langswapening

0742 longitudinal section
f section longitudinale
d Längsschnitt
e sección longitudinal
n langsdoorsnede

0743 longitudinal weld
f soudure longitudinale
d Längsschweissung
e soldadura longitudinal
n langslas

0744 long-term test
f essai à long terme
d Langzeitversuch
e ensayo de larga duración
n lange-duur proef

0745 loop anchorage
f ancrage bouclé
d Schlaufenverankerung
e anclaje de bucle
n lusverankering

0746 loss of head
f perte de charge
d Fallhöhenverlust
e pérdida de carga
n drukverlies

0747 loss of pressure
f chute de pression
d Druckverlust
e pérdida de presión
n drukverlies

0748 loss of stress
f perte de contrainte
d Spannungsverlust
e pérdida de tensión
n spanningsverlies

M

0749 magnesium cement
f ciment de magnésie
d Magnesiumzement
e cemento de magnesia
n magnesium cement

0750 main beam
f sous-poutre principale
d Unterzug
e viga principal; jácena
n moerbalk

0751 main girder
f poutre principale
d Hauptträger
e viga principal; jácena
n hoofdligger

0752 main reinforcement
f armature principale
d Hauptbewehrung
e armadura principal
n hoofdwapening

0753 maintenance
f entretien
d Unterhalt
e conservación; entretenimiento
n onderhoud

0754 mandrel (of bar bending machine)
f mandrin (d'une cintreuse)
d Biegedorn
e mandril (de máquina de doblado de barras)
n doorn (van buigmachine)

0755 man-hour
f heure de main-d'oeuvre
d Arbeitsstunde
e hora-hombre
n man-uur

0756 manpower
f main-d'oeuvre
d Arbeitskraft
e mano de obra
n mankracht; arbeidskracht(en)

0757 manufacture *v*
f fabriquer
d herstellen
e fabricar; manufacturar
n fabriceren

0758 mark(ing)
f marque; repère
d Merkzeichen
e marca; señal; referencia
n merkteken

0759 marl
f marne
d Mergel
e marga
n mergel

0760 martempering
f trempe en bain chaud
d Warmbadhärten
e martempering
n mertemperen

0761 martensite
f martensite
d Martensit
e martensita
n martensiet

0762 material
f matière
d Material
e material
n materiaal

0763 material retained on sieve
f refus au tamis
d Siebrückstand
e material retenido por el tamíz
n zeefrest (fractie)

* **mean** → 0067

0764 measure
f mesure
d Masstab
e medida; tamaño
n maatstaf

0765 measuring box
f boîte de mesure
d Messbüchse
e caja de medida
n meetbak

0766 measuring cylinder
f éprouvette graduée
d Messzylinder
e recipiente de medida; vaso de medida
n maatglas

0767 mechanical testing
f essai mécanique
d mechanische Prüfung
e ensayo mecánico
n mechanisch onderzoek

0768 mechanics
f mécanique
d Mechanik
e mecánica
n mechanica

0769 membrane
f membrane
d Membran
e membrana
n membraan

0770 membrane stress
f contrainte de membrane
d Membranspannung
e tensión de membrana
n membraamspanning

0771 meridian
f méridien
d Meridian
e meridiano
n meridiaan

0772 meridian stress
f contrainte méridienne
d Meridianspannung
e tensión según los meridianos
n meridiaanspanning

0773 mesh
f maille
d Masche
e malla
n maas

0774 mesh reinforcement
f armature en treillis
d Netzbewehrung
e armadura de malla
n netwapening

0775 metacentre
f métacentre
d Metazentrum
e metacentro
n metacenter

0776 metal
f métal
d Metall
e metal
n metaal

0777 metal bearing
f appui métallique
d Metallager
e apoyo metálico
n metalen oplegging

0778 micro-crack
f microfissure
d Mikroriss
e microfisura
n microscheur

0779 middle strip (flat-slab floor)
f bande de travée (dalle-champignon)
d Feldstreifen (Pilzdecke)
e abaco central
n middenstrook (paddestoelvloer)

0780 mid-span moment
f moment en travée
d Feldmoment
e momento en el centro de tramo
n veldmoment

0781 mid-span reinforcement
f armature en travée
d Feldbewehrung
e armadura en el centro del tramo
n veldwapening

0782 mineral
f minéral
d Mineral
e mineral
n mineraal

0783 mix *v*
f mélanger; malaxer
d mischen
e mezclar
n mengen

0784 mix; mixture
f mélange
d Mischung
e mezcla
n mengsel

0785 mixing drum
f tambour mélangeur
d Mischtrommel
e tambor mezclador
n mengtrommel

0786 mixing plant
f centrale de mélange
d Mischanlage
e central de hormigonado
n menginstallatie

0787 mixing screw
f vis mélangeuse
d Mischschnecke
e tornillo mezclador
n mengschroef

0788 mixing water
f eau de gâchage
d Anmachwasser
e agua de amasado
n aanmaakwater

* **mixture** → 0784

0789 model testing
f essai sur modèle
d Modellversuch
e ensayo sobre modelo
n modelonderzoek

0790 modular ratio method
f méthode élastique de calcul
d n-Verfahren
e método de relación modular; método n
n n-methode

0791 modulus
f module
d Modul
e módulo
n modulus

0792 modulus of elasticity
f module d'élasticité
d Elastizitätsmodul
e módulo de elasticidad
n elasticiteitsmodulus

* **modulus of rigidity** → 1052

0793 moisture
f humidité
d Feuchtigkeit
e humedad
n vochtigheid

0794 moment
f moment
d Moment
e momento
n moment

0795 moment coefficient
f coefficient de moment
d Momentenbeiwert
e coeficiente de momentos
n momentcoëfficient

0796 moment diagram
f diagramme de moments
d Momentenlinie
e diagrama de momentos
n momentenlijn

0797 moment of inertia
f moment d'inertie
d Trägheitsmoment
e momento de inercia
n traagheidsmoment

0798 monolithic
f monolithique
d monolitisch
e monolítico
n monoliet

0799 monorail
f monorail
d Einschienenbahn
e monorail
n monorail

0800 mortar
f mortier
d Mörtel
e mortero
n mortel

0801 mould *v*
f mouler
d formen
e formar; moldear; modelar
n vormen

0802 mould
f moule
d Form
e molde
n mal; model

* **mould oil** → 0549

0803 moving load
f charge mobile
d Wanderlast
e carga móvil
n mobiele belasting

0804 mushroom floor; drop
f plancher champignon; abaque
d Pilzdecke; Stützenkopfplatte
e forjado fungiforme; placa fungiforme; abaco
n paddestoelvloer; kolomplaat

N

0805 needle
f aiguille
d Nadel
e aguja
n naald

0806 neutral axis
f axe neutre
d neutrale Achse; Nullinie
e eje neutro
n neutrale lijn

0807 non-destructive testing
f essai non destructif
d zerstörungsfreie Prüfung
e ensayo no destructivo
n niet-destructief onderzoek

0808 normal force diagram
f diagramme d'effort normal
d Normalkraftdiagram
e diagrama de esfuerzos normales
n normaalkrachtenlijn

0809 normal stress
f contrainte normale
d Normalspannung
e tensión normal
n normaalspanning

0810 north-light roof
f toiture en shed
d Sheddach
e cubierta en diente de sierra
n sheddak

0811 "no-slump" concrete
f béton sec
d erdfeuchter Beton
e hormigón "sin asiento"; hormigón seco
n aardvochtig beton

0812 notch
f entaille
d Kerbe
e muesca
n kerf

0813 notch effect
f effet d'entaille
d Kerbwirkung
e efecto de entalladura
n kerfwerking

0814 notch value; impact value
f résilience
d Kerbschlagwert
e resiliencia
n kerfslagwaarde

0815 numerical
f numérique
d numerisch
e numérico
n numeriek

0816 nut
f écrou
d Mutter
e tuerca; rosca
n moer

0817 nut anchorage
f ancrage à écrou
d Verankerung mit Mutter
e anclaje de rosca
n moerverankering

O

0818 oblique
f oblique
d schräg
e oblícuo
n scheef

0819 oil
f huile
d Öl
e aceite
n olie

0820 order
f commande
d Auftrag
e orden; pedido
n opdracht; order

0821 orthogonal
f orthogonal
d orthogonal
e ortogonal
n orthogonaal

0822 orthotropic
f orthotrope
d orthotrop
e ortótropo
n orthotroop

0823 oscillate *v*
f osciller
d schwingen
e oscilar
n slingeren

* **outside** → 0498

0824 oval-section steel
f acier de section ovale
d Stahl mit ovalem Querschnitt
e acero de sección oval
n staal van ovale doorsnede

0825 oval wire
f fil ovale
d ovaler Draht
e alambre oval
n ovaal draad

0826 overload *v*
f surcharger
d überbelasten
e sobrecargar
n overbelasten

0827 overpass
f passage supérieur
d Überführung
e paso superior; viaducto
n viaduct

0828 oxy-acetylene welding
f soudage oxy-acétylénique
d autogene Schweissung
e soldadura oxi-acetilena
n autogeen lassen

P

0829 pack *v* (horizontal joint with mortar to form a bedding)
f sceller (un joint horizontal au mortier)
d unterstopfen (einer Horizontalfuge mit Mörtel)
e almohadillar (junta horizontal con cama de mortero)
n ondersabelen

0830 packing ring
f segment (de piston)
d Dichtring; Manschette
e empaquetadura
n pakkingring; manchet

0831 panel
f panneau
d Platte
e panel
n paneel

0832 parabola
f parabole
d Parabel
e parábola
n parabool

0833 parameter
f paramètre
d Parameter
e parámetro
n parameter

0834 part by volume
f proportion en volume
d Raumteil
e parte del volúmen
n volumedeel

0835 part by weight
f proportion en poids
d Gewichtsteil
e parte del peso
n gewichtsdeel

0836 partial
f partiel
d teilweise; partiell
e parcial
n partiëel

0837 partial prestress(ing)
f précontrainte partielle
d teilweise Vorspannung
e pretensión parcial
n gedeeltelijke voorspanning

0838 particle size
f grosseur de grain; calibre
d Korngrösse
e tamaño de la partícula
n korrelgrootte

0839 particle size analysis
f analyse granulométrique
d Korngrössenbestimmung
e análisis del tamaño de la partícula
n granulometrisch onderzoek

0840 particle strength
f résistance de grains
d Kornfestigkeit
e resistencia de la partícula
n korrelsterkte

0841 patent
f brevet
d Patent
e patente
n patent; octrooi

0842 patent *v* (of steel)
f patenter (de l'acier)
d patentieren (von Stahl)
e patentar (del acero)
n patenteren (van staal)

0843 pavement; road pavement
f revêtement (de chaussée)
d Strassendecke
e empedrado
n rijdek

0844 pea gravel
f gravier roulé
d Perlkies; Erbskies
e gravilla
n parelgrind

0845 pearlite
f perlite
d Perlit
e perlita
n perliet

0846 penetration of water
f pénétration d'eau
d Wassereindringung
e penetración de agua
n waterindringing

0847 penetrometer
f pénétromètre
d Eindringungsmesser
e penetrómetro
n sondeerapparaat

0848 percentage by volume
f pourcentage en volume
d Volumenprozent
e cuantía volumétrica
n volumepercentage

0849 percentage by weight
f pourcentage en poids
d Gewichtsprozent
e cuantía gravimétrica
n gewichtspercentage

* **perimeter** → 0221

0850 permissible stress
f contrainte admissible
d zulässige Spannung
e tensión admisible
n toelaatbare spanning

0851 photo-elastic test
f essai photoélastique
d spannungsoptische Prüfung
e ensayo fotoelástico
n foto-elastisch onderzoek

0852 pier of bridge
f pile de pont
d Pfeiler einer Brücke
e pila de puente
n pijler van een brug

0853 pile (for foundation)
f pieu (de fondation)
d Pfahl (für Gründung)
e pilote
n paal (voor fundering)

0854 pile head
f tête de pieu
d Pfahlkopf
e cabeza del pilote
n paalkop

0855 pile shaft
f fût de pieu
d Pfahlschaft
e fuste del pilote
n paalschacht

0856 pile tip
f pointe de pieu
d Pfahlspitze
e punta del pilote; pie del pilote
n paalpunt

0857 pipe
f tuyau
d Rohr
e tubería; tubo
n pijp; buis

0858 pitch (of a screw thread; of a spiral)
f pas (hélice; filetage)
d Steigung; Ganghöhe (eines Gewindes oder einer Spirale)
e paso (de la rosca de un tornilla; de una espiral)
n spoed (van een schroefdraad; van een spiraal)

0859 place *v*
f placer; mettre en oeuvre
d einbauen; einbringen
e colocar; verter
n storten

0860 plain; unreinforced
f non armé
d unbewehrt
e llano; liso
n ongewapend

0861 plain bars
f armatures lisses
d glatter Bewehrungsstahl
e barras lisas
n glad wapeningsstaal

* **plain girder** → 1115

0862 plain wire
f fil lisse
d glatter Draht
e alambre liso
n glad draad

0863 plane
f plan
d Ebene
e plano
n vlak

0864 plaster *v*
f plâtrer
d putzen
e yesar
n pleisteren

0865 plastic; plastics
f matière plastique
d Kunststoff
e plástico
n kunststof

0866 plastic deformation
f déformation plastique
d plastische Verformung
e deformación plástica
n plastische vervorming

0867 plastic shrinkage
f retrait plastique
d plastische Schwindung
e retracción plástica
n plastische krimp

0868 plastic strain
f déformation plastique
d plastische Dehnung
e deformación plástica unitaria
n plastische vervorming

0869 plasticiser
f plastifiant
d Betonverflüssiger
e plastificante
n plastificeerder

0870 plasticity (behaviour of materials)
f plasticité (comportement de matières)
d Plastizität (Materialverhalten)
e plasticidad (comportamiento de los materiales)
n plasticiteit (materiaalgedrag)

0871 plasticity (consistency)
f plasticité (consistance)
d Plastizität (Konsistenz)
e plasticidad (consistencia)
n plasticiteit (consistentie)

* **plastics** → 0865

0872 plate
f plaque
d Platte; Scheibe
e placa
n plaat

0873 plate action
f effet de plaque
d Plattenwirkung
e trabajo de placa
n schijfwerking

* **plate iron** → 1059

0874 play
f jeu
d Spiel
e juego; huelgo
n speling

0875 plug
f tampon; bouchon
d Pfropfen
e tapón
n prop

0876 plumb-bob
f fil à plomb
d Senkblei; Schnurlot
e plomada
n schietlood

0877 plunger
f piston plongeur
d Tauchkolben
e pistón
n plunjer

0878 plunger pump
f pompe à piston plongeur
d Tauchkolbenpumpe
e pistón de bomba; émbolo de bomba
n plunjerpomp

* **point load** → 0268

0879 point of zero moment
f point de moment nul
d Momentennullpunkt
e punto de momento nulo
n momentennulpunt

0880 polish *v*
f polir
d polieren
e pulir
n polijsten

0881 polished surface
f surface polie
d polierte Oberfläche
e superficie pulida
n gepolijst oppervlak

* **pollute** *v* → 0306

* **pollution** → 0307

0882 population
f population
d Bevölkerung
e población
n populatie

* **pore space** → 1350

* **pore volume** → 1350

* **pores** → 0026

* **pores** → 1348

* **pores content** → 1349

0883 porous
f poreux
d porös
e poroso
n poreus

0884 portal crane
f grue portique
d Portalkran
e grúa-pórtico
n portaalkraan

0885 portal frame
f portique
d Portalrahmen
e pórtico
n portaal

0886 position
f position
d Lage
e posición; sitio
n ligging

0887 post-tensioning
f postcontrainte
d Vorspannen (nach dem Erhärten des Betons)
e postesado (pretensado con armaduras postesas)
n voorspanning met nagerekt staal

0888 pot-life
f délai avant prise (résine)
d Topfzeit
e tiempo útil de utilización
n houdbaarheid (na mengen)

0889 pour *v*
f mettre en oeuvre
d einbringen
e moldear; verter
n storten

0890 power shovel
f pelle mécanique
d Hochlöffelbagger
e excavadora; pala mecánica
n mechanische schop

0891 precast *v*; **prefabricate** *v*
f préfabriquer
d vorfertigen
e prefabricar
n prefabriceren

0892 precast concrete
f béton préfabriqué
d Betonfertigteile
e hormigón prefabricado
n geprefabriceerd beton; prefabbeton

* **prefabricate** *v* → 0891

0893 preflex girder
f poutre préflex
d Preflex-Balken
e viga preflex
n preflexligger

0894 preliminary design
f avant-projet
d Vorentwurf
e anteproyecto
n voor-ontwerp

0895 preliminary test
f essai préliminaire
d Eignungsprüfung
e ensayo preliminar; ensayo previo
n geschiktheidsproef

0896 preserve *v*
f conserver
d konservieren
e preservar; proteger; conservar
n verduurzamen; conserveren

* **pressure** → 0258

0897 pressure cell
f cellule manométrique
d Druckdose
e célula de presión
n drukdoos

0898 pressure gauge
f manomètre
d Manometer
e manómetro
n manometer

0899 prestress *v*
f mettre en précontrainte
d vorspannen
e pretensar
n voorspannen

0900 prestress
f précontrainte
d Vorspannung
e pretensión
n voorspanning

0901 prestressed
f précontraint
d vorgespannt
e pretensado
n voorgespannen

0902 prestressing bed (long)
f banc de précontrainte (grand banc)
d Spannbett (lang)
e banco de tesado (largo); bancada de pretensado (largo)
n spanbaan; lange bank

0903 prestressing bed (short) (individual mould)
f banc de précontrainte (moule individuel)
d Spannbett (kurz)
e banco de tesado (corto); bancada de pretensado (corto)
n spanraam

0904 prestressing by winding (merry-go-round method)
f précontrainte par enroulement
d Vorspannung durch Wickeln (Wickelverfahren)
e pretensado por enrollamiento
n ringvoorspanning (merry-go-round)

0905 prestressing force
f force de précontrainte
d Vorspannkraft
e fuerza de pretensado
n voorspankracht

0906 prestressing steel
f acier de précontrainte
d Spannstahl
e acero de pretensado
n voorspanstaal

0907 prestressing system
f système de précontrainte
d Vorspannverfahren
e sistema de pretensado
n spansysteem

0908 pretensioning
f précontrainte par prétension
d Spannbettvorspannung
e pretensado mediante armaduras pretesas
n voorspanning met voorgerekt staal

0909 principal stress
f contrainte principale
d Hauptspannung
e tensión principal
n hoofdspanning

0910 principal tensile stress
f contrainte principale de traction
d Hauptzugspannung
e tensión principal de tracción
n hoofdtrekspanning

0911 prismatic
f prismatique
d prismatisch
e prismático
n prismatisch

0912 probability theory
f théorie probabiliste
d Wahrscheinlichkeitstheorie
e teoría probabilística
n waarschijnlijkheidsleer

0913 production line
f ligne de production
d Fertigungsstrasse
e línea de fabricación
n produktielijn

0914 profiling
f profilage
d Profilierung
e perfilando
n profilering

* **project** → 1022

* **promotor** → 0150

0915 prop *v*
f étayer
d stützen; abstützen
e apuntalar
n schoren

0916 prop; strut
f étai; étançon
d Stütze; Gerüstständer
e puntal; biela
n schoor; stut

0917 proportion *v*
f doser
d dosieren
e dosificar
n doseren

0918 protruding bar
f barre saillante
d herausragender Stab
e barra saliente
n stekeind

0919 pulsating stress
f contrainte pulsatoire
d Schwellspannung
e tensión oscilante
n rimpelspanning

0920 pumice concrete
f béton de ponce
d Bimsbeton
e hormigón de pumita
n bimsbeton

0921 pump up *v*
f pomper
d aufpumpen
e bombear
n oppompen

0922 punch *v*
f poinçonner
d stanzen; durchstanzen
e punzonar
n ponsen

0923 punching shear stress
f contrainte de poinçonnement
d Durchstanz-Schubspannung
e tensión de punzonamiento
n ponsspanning

0924 purlin
f panne
d Pfette
e vigueta; correa
n gording

0925 push down (a prestressing wire)
f dévier en bas (un fil de précontrainte)
d niederdrücken (eines spanndrahtes)
e empujar hacia abajo (un alambre de pretensado)
n neerdrukken (van een spandraad)

Q

* **quality class** → 0571

0926 quartz powder
f farine de quartz
d Quarzmehl
e polvo de cuarzo
n kwartsmeel

0927 quay structure
f construction de quai
d Kaianlage
e estructura de muelle de atraque
n kadeconstructie

0928 quay wall
f mur de quai
d Kaimauer
e muro de muelle
n kademuur

R

0929 radial
f radial
d radial
e radial
n radiaal

0930 radial stress
f contrainte radiale
d Radialspannung
e tensión radial
n radiaalspanning

0931 radius of curvature
f rayon de courbure
d Krümmungsradius
e radio de curvatura
n kromtestraal

0932 radius of inertia
f rayon d'inertie
d Trägheitsradius
e radio de inercia
n traagheidsstraal

0933 rainwater
f eau de pluie
d Regenwasser
e agua lluvia
n hemelwater

0934 raking pile; batter pile
f pieu incliné
d Schräggründungspfahl; Schrägpfahl
e pilote inclinado
n schoorpaal

0935 raking position
f inclinaison
d Schräglage
e posición inclinada
n schoorstand

0936 ramp
f rampe
d Rampe
e rampa; pendiente
n helling; oprit

0937 random test
f essai au hasard
d Stichprobe
e ensayo aleatorio
n steekproef

0938 rapid-hardening cement
f ciment à durcissement rapide
d frühhochfester Zement
e cemento de endurecimiento rápido
n snel verhardend cement

0939 rate of loading
f taux de chargement
d Belastungsgeschwindigkeit
e velocidad de carga
n belastingssnelheid

0940 ready-mixed concrete
f béton prêt à l'emploi
d Transportbeton
e hormigón preamasado
n transportbeton

0941 ready-mixed concrete plant
f centrale de béton prêt à l'emploi
d Transportbeton-Aufbereitungsanlage
e fábrica de hormigón preparado
n betonmortelcentrale

0942 recess
f évidement
d Aussparung
e vacío; cavidad
n sparing

0943 reduction
f réduction
d Verminderung
e reducción
n reductie

0944 reel
f bobine; rouleau
d Spule; Rolle
e torno; polea; cabrestante
n klos

0945 refusal (pile driven to refusal)
f refus (pieu battu au refus)
d Fusstützung (eines Rammpfahls)
e rechazo
n stuit (paal op stuit geheid)

0946 regulation(s)
f règlement
d Vorschrift(en)
e reglamento
n voorschrift

0947 reinforce *v*
f armer
d bewehren
e armar
n wapenen

0948 reinforced concrete
f béton armé
d Stahlbeton
e hormigón armado
n gewapend beton

0949 reinforcement
f armature
d Bewehrung
e armadura
n wapening

0950 reinforcement cage
f cage d'armatures
d Bewehrungskorb
e jaula de armaduras
n wapeningskooi; wapeningskorf

0951 reinforcement cutting shears
f cisaille pour acier d'armature
d Betonstahlschneider
e cizallas para armadura
n betonijzerschaar

0952 reinforcement detector
f détecteur d'armatures
d Bewehrungsdetektor
e detector de armaduras
n wapeningsdetector

0953 reinforcement percentage
f pourcentage d'armatures
d Bewehrungsprozentsatz
e cuantía de armadura
n wapeningspercentage

0954 reinforcing bar
f barre d'armature
d Bewehrungsstab
e barra de armadura
n wapeningsstaaf

0955 reinforcing steel
f acier d'armature
d Betonstahl; Bewehrungsstahl
e acero para armadura
n wapeningsstaal; betonstaal

0956 relative humidity
f humidité relative
d relative Feuchtigkeit
e humedad relativa
n relatieve vochtigheid

0957 relaxation
f relaxation
d Relaxation
e relajación
n relaxatie

0958 release *v*
f relâcher
d entspannen
e soltar; aflojar
n ontspannen

0959 release *v* (tension)
f détendre
d ablassen (der Spannung)
e destesar
n aflaten (van spanning)

0960 release *v* (a wedge)
f débloquer (une clavette)
d lösen (eines Keils)
e soltar (una cuña); aflojar (una cuña)
n lossen (van een wig)

0961 relieve *v*
f décharger
d entlasten
e rebajar
n ontlasten

0962 research
f recherche
d Forschung
e investigación
n research; speurwerk

0963 reservoir
f réservoir
d Behälter
e depósito
n reservoir

0964 resilient joint
f joint élastique
d elastische Fuge
e junta elástica
n elastische voeg

0965 resistance
f résistance
d Widerstand
e resistencia
n weerstand

0966 resistance butt weld
f soudure par résistance pure
d Widerstandsstumpfnaht
e soldadura a tope por resistencia
n weerstandsstuiklas

0967 resistance welding
f soudage par résistance
d Widerstandsschweissung
e soldadura por resistencia
n weerstandslassen

0968 resistant
f résistant
d beständig
e resistente
n bestendig

0969 resultant
f résultante
d resultierende
e resultante
n resultante

0970 retaining wall
f mur de soutènement
d Stützmauer
e muro de contención
n keermuur

0971 retard *v*
f retarder
d verzögern
e retardar; aplazar
n vertragen

0972 retension *v*
f retendre
d nachspannen
e retesar
n naspannen

0973 reverse bend test
f essai de flexion alternée
d Hin- und Herbiegeversuch
e ensayo de doblado alternativo
n terugbuigproef

0974 reversible
f réversible
d umkehrbar
e reversible
n reversibel

0975 rib
f nervure
d Rippe
e costilla; nervio
n rib

0976 ribbed floor
f plancher à nervures
d Rippendecke
e forjado nervado
n ribbenvloer

0977 ribbed slab
f dalle nervurée
d Rippenplatte
e placa nervada
n geribde plaat

0978 rich bottom layer (as first pour in the formwork)
f couche de béton riche (coulé sous l'eau)
d Anschlussmischung (die erste in die Schalung eingebrachte Schicht)
e capa inferior rica (por ser la primera vertida en el encofrado)
n smeerbed

0979 rich concrete
f béton riche
d fetter Beton
e hormigón rico
n vet beton

0980 rigid joint
f joint rigide
d starre Fuge
e junta rígida
n starre voeg

0981 rigidity
f rigidité
d Steifigkeit
e rigidez
n stijfheid

0982 ring beam
f poutre annulaire
d Ringbalken
e viga circular
n ringbalk

0983 rise (of an arch)
f flèche (d'un arc)
d Stich (eines Bogens)
e rebajamiento (de un arco); flecha (de un arco)
n pijl (van een boog)

0984 river gravel
f gravier de rivière
d Flusskies
e guijarro; grava de río; grava rodada
n riviergrind

0985 river sand
f sable de rivière
d Flussand
e areno de río
n rivierzand

* **road pavement** → 0843

0986 rock anchor
f ancre de rocher
d Felsanker
e ancla en roca
n rotsanker

0987 rocker
f bascule; balancier
d Kipphebel
e rótula
n pendel

0988 rocker bearing
f appui à bascule
d Kipplager
e apoyo articulado
n pendeloplegging

0989 rocker-dump hand-cart
f benne equilibrée
d Japaner
e cuba volquete de mano
n japanner

0990 roll *v*
f laminer
d walzen
e rodear; cilindrar; laminar
n walsen

0991 roll(er)
f rouleau
d Walze
e rodillo; apisonadora
n rol

0992 roller bearing
f appui à rouleau
d Rollenlager
e apoyo de rodillos
n roloplegging

0993 rolled wire
f fil laminé
d Walzdraht
e alambra laminado
n gewalst draad

0994 rolling skin
f peau de laminage
d Walzhaut
e capa de rodadura
n walshuid

0995 roof
f toit; toiture
d Dach
e cubierta; techumbre
n dak

0996 roof beam
f poutre de couverture
d Dachbalken
e viga de cubierta
n dakbalk

0997 roof frame; roof truss
f ferme de toiture
d Dachbinder
e correa de cubierta
n dakspant

0998 roof load(ing)
f charge de toiture

d Dachbelastung
e carga de cubierta
n dakbelasting

0999 roof structure
f construction de la toiture
d Dachkonstruktion
e estructura de cubierta
n dakconstructie

* **roof truss** → 0997

1000 rotation
f rotation
d Drehung
e giro; rotación
n rotatie

1001 round bar
f barre ronde
d Rundstab
e barra redonda; redondo
n rondstaal

1002 round-hole screen
f passoire (à trous ronds)
d Rundlochsieb
e tamíz de malla redonda
n plaatzeef

1003 rounded aggregate
f agrégat roulé
d runder Zuschlag
e árido rodado
n rond toeslagmateriaal

1004 rubber bearing
f appui en caoutchouc
d Gummilager
e apoyo de caucho
n rubberoplegging

1005 rubber pad
f plaque en caoutchouc
d Gummikissen
e almohadilla de caucho
n rubberblok

1006 runway
f piste d'aviation
d Startbahn; Landebahn; Piste
e pista de aeropuerto
n startbaan; landingsbaan

1007 rust
f rouille
d Rost
e orín; herrumbre; óxido
n roest

S

1008 saddle
f selle
d Sattel
e soporte; silla
n zadel

1009 safety
f sécurité
d Sicherheit
e seguridad
n veiligheid

1010 safety against cracking
f sécurité vis-à-vis de la fissuration
d Rissicherheit
e seguridad a fisuración
n scheurveiligheid

1011 safety against failure
f sécurité vis-à-vis de la rupture
d Bruchsicherheit
e seguridad a rotura
n breukveiligheid

1012 sample
f échantillon
d Muster
e muestra
n monster

1013 sand
f sable
d Sand
e arena
n zand

1014 sand pot
f boîte à sable
d Sandtopf
e caja de arena
n zandpot

1015 sandblast *v*
f sable
d sandstrahlen
e chorrear de arena
n zandstralen

1016 sandwich plate
f plaque sandwich
d Sandwichplatte
e placa sandwich
n sandwichplaat

1017 saturated
f saturé
d gesättigt
e saturado
n verzadigd

1018 scaffold(ing)
f échafaudage
d Gerüst; Rüstung
e malecón; muelle; escollera
n steiger (als constructie)

* **scale** → 0075

1019 scale (dimensional)
f échelle (des dimensions)
d Masstab
e escala (dimensional)
n schaal (afmeting)

1020 scale off *v*
f écailler
d abblättern; abschuppen
e descantillarse
n afschilferen

1021 scatter
f dispersion
d Streuung
e dispersión
n spreiding

1022 scheme; project
f projet
d Projekt; Bauvorhaben
e proyecto
n project

1023 scour *v*
f affouiller
d unterspülen
e socavar; arrastrar
n uitkolken

1024 screed board
f latte de réglage
d Glättbohle
e escantillón; rasero
n afrijbalk

1025 screen
f crible; tamis
d Sieb; Gittersieb
e criba; cedazo; tamíz
n zeef (grof)

1026 screen size; sieve size
f numéro du tamis
d Siebgrösse
e tamaño del tamíz
n zeefmaat

1027 screw anchorage
f ancrage à vis
d Schraubverankerung
e anclaje de tornillo
n schroefverankering

* **screw conveyor** → 1410

1028 screw jack
f vérin à vis
d Schraubenwinde; Schraubenspindel; Spindel
e gato de tornillo
n schroefvijzel

1029 screw thread
f filetage de vis
d Schraub(en)gewinde
e rosca del tornillo
n schroefdraad

1030 screwed sleeve
f manchon à vis
d Gewindehülse
e manguito roscado
n schroefhuls

1031 scriber
f pointe à tracer
d Kratzfeder
e marcador
n kraspen

1032 scrubbed concrete surface
f parement en béton lavé
d Waschbeton-Oberfläche
e superficie lavada (dejando al descubierto al árido del hormigón)
n uitgewassen betonoppervlak

1033 sea gravel
f gravier de mer
d Seekies
e grava de mar
n zeegrind

* **seal** → 0563

1034 seal *v*
f sceller; étancher
d dichten; abdichten
e obturar; cerrar; sellar
n afdichten

* **sealing coat** → 0126

1035 sealing compound
f mastic d'étanchéité
d Dichtungsmasse
e producto de selladura; mástique
n dichtingsmiddel; kit

1036 secondary stress
f contrainte secondaire
d Nebenspannung
e tensión secundaria
n secundaire spanning

1037 section
f section
d Schnitt
e sección
n doorsnede

1038 section modulus
f module de résistance
d Widerstandsmoment
e momento resistente de la sección
n weerstandsmoment

1039 section(al) steel
f acier profilé
d Profilstahl
e acero perfilado
n profielstaal

1040 segregation
f ségrégation
d Entmischung
e segregación; separación
n ontmenging

1041 set out *v* (dimensions)
f piqueter (dimensions)
d abstecken (Masse)
e fijar (dimensiones); marcar (dimensiones)
n uitzetten (maatvoering)

1042 set-screw
f vis de réglage
d Stellschraube

e tornillo de ajuste; tornillo prisionero
n stelschroef

1043 setting
f prise
d Erstarren; Abbinden
e fraguado
n binding

1044 setting accelerator
f accélérateur de prise
d Erstarrungsbeschleuniger
e acelerador de fraguado; acelerante de fraguado
n bindingsversneller

1045 setting retarder
f retardeur de prise
d Erstarrungsverzögerer
e retardador de fraguado
n bindingsvertrager

1046 settlement
f tassement
d Setzung; Senkung
e asiento (de un terreno)
n zetting (zakking)

1047 sewer pipe
f conduite d'égout
d Abwasserrohr
e tubería de desagüe; tubería de saneamiento; alcantarilla
n rioolbuis

1048 shaking screen
f crible à secousses
d Schüttelsieb
e tamíz de sacudidas; tamíz oscilante
n schudzeef

1049 shatter *v*
f briser
d zerschlagen
e triturar; machacar; astillar
n verbrijzelen

1050 shear force
f effort tranchant
d Querkraft
e esfuerzo cortante
n dwarskracht

1051 shear force diagram
f diagramme d'effort tranchant
d Schwerkraftlinie
e diagrama de esfuerzos cortantes
n dwarskrachtenlijn

1052 shear modulus; modulus of rigidity
f module de cisaillement
d Schubmodul
e módulo de rigidez
n glijdingsmodulus

1053 shear reinforcement
f armature de cisaillement
d Schubbewehrung
e armadura transversal
n schuifwapening

1054 shear stress
f contrainte de cisaillement
d Schubspannung
e tensión cortante
n schuifspanning

1055 shear strength
f résistance au cisaillement
d Schubfestigkeit
e resistencia a cizallamiento
n schuifsterkte

1056 shearing
f cisaillement
d Schub
e cizallamiento
n afschuiving

1057 sheath (of cable)
f gaine (d'un câble)
d Hüllrohr (eines Kabels)
e vaina; funda
n omhullingsbuis; aalhuid

1058 sheet (plastic); **foil** (metal)
f feuille (matière plastic; métal)
d Folie (Kunststoff; Metall)
e lámina; hoja
n folie (kunststof; metaal)

1059 sheet iron; plate iron
f tôle de fer
d Eisenblech
e palastro
n plaatijzer

1060 sheet pile
f palplanche
d Spundbohle
e tablestaca
n damplank

1061 sheet pile wall; sheet piling
f rideau de palplanches
d Spundwand
e tablestacado
n damwand

* **sheet piling** → 1061

1062 shell (structural)
f coque (construction)
d Schale (Bauteil)
e lámina (estructural)
n schaal (constructie)

1063 shore *v*
f étrésillonner
d absteifen
e apuntalar
n schoren

1064 shore
f étrésillon
d Strebe; Schrägsteile
e borde; orilla; puntal
n schoor

1065 shorten *v*
f raccourcir
d verkürzen
e acortar
n verkorten

1066 short-term test
f essai de courte durée
d Kurzzeitversuch
e ensayo de corta duración
n korte-duur proef

1067 shotcrete
f béton projeté
d Spritzbeton
e hormigón lanzado; gunita
n spuitbeton

1068 shrinkage
f retrait
d Schwinden
e retracción
n krimp

1069 shrinkage cracks
f fissures de retrait
d Schwindrisse
e fisuras de retracción
n krimpscheuren

1070 shrinkage factor
f facteur de retrait
d Schwindfaktor
e coeficiente de retracción; factor de retracción
n krimpfactor

1071 shrinkage gradient
f gradient de retrait
d Schwindungsgefälle
e gradiente de retracción
n krimpgradiënt

1072 shrinkage joint
f joint de retrait
d Scheinfuge
e junta de retracción
n krimpvoeg

1073 shrinkage modulus
f module de retrait
d Schwindmodul
e módulo de retracción
n krimpmodulus

1074 shrinkage reinforcement
f armature de retrait
d Schwindbewehrung
e armadura de retracción
n krimpnet; krimpwapening

1075 shrinkage stress
f contrainte de retrait
d Schwindspannung
e tensión de retracción
n krimpspanning

1076 shutter *v*; **erect framework** *v*
f coffrer
d einschalen
e tablear
n bekisten

* **shuttering** → 0548

1077 sieve
f tamis
d Sieb

e tamíz; criba; cedazo
n zeef

1078 sieve analysis
f analyse granulométrique
d Siebanalyse; Siebversuch
e análisis granulométrico
n zeefanalyse

1079 sieve residue
f refus de tamisage
d Siebrückstand
e residuo en el tamíz; porcentaje retenido en el tamíz
n zeefrest

* **sieve size** → 1026

1080 silo; bin
f silo
d Silo
e depósito; silo
n silo

1081 silo effect
f effet de silo
d Silowirkung
e efecto pared
n silowerking

1082 sintering
f sintérisation
d Sinterung
e sinterización
n sinteren

1083 site; construction site; job site
f chantier
d Baustelle
e ubicación; sitio; situación
n bouwplaats; bouwterrein; werkterrein

1084 site-mixed concrete
f béton fabriqué sur le chantier
d auf der Baustelle gemischter Beton
e hormigón fabricado in situ
n op de bouwplaats vervaardigd beton

1085 skew
f biais
d schief
e inclinado; oblícuo; esviado
n scheluw; scheef

1086 skew slab
f dalle biaise
d schiefe Platte
e placa esviada
n scheve plaat

1087 skew slab bridge
f pont-dalle biais
d schiefe Plattenbrücke
e tablero de puente en esviaje
n scheve plaatbrug

1088 skilled workman
f ouvrier spécialisé
d gelernter Arbeiter; Facharbeiter
e obrero especializado; operario
n geschoold arbeider

1089 skip
f benne
d Förderkübel
e caja; cajón; cuba
n stortbak; kubel

1090 slab
f dalle
d Platte
e placa; losa
n plaat

1091 slab bridge
f pont-dalle
d Plattenbrücke
e puente-losa
n plaatbrug

1092 slab floor
f plancher-dalle
d Plattendecke
e placa de forjado
n plaatvloer

1093 slack
f lâche; détendu
d Schlaff
e flojo; aflojado
n slap

1094 slate
f ardoise
d Schiefer

e pizarra
n lei

* **sleeper** (railroad) → 1273

1095 sleeve
f manchon
d Muffe
e manguito
n manchet

1096 slenderness
f élancement
d Schlankheit
e esbeltez
n slankheid

1097 slide *v*
f glisser
d gleiten; rutschen
e deslizar
n glijden

1098 slide *v* **into one another**
f faire entre l'un dans l'autre
d ineinanderschieben
e deslizar uno sobre otro
n ineenschuiven

1099 slide rule
f règle à calcul
d Rechenschieber
e regla de cálculo
n rekenlineaal

1100 sliding formwork; slipforms
f coffrage glissant
d Gleitschalung
e encofrado deslizante
n glijbekisting

1101 sliding reaction
f réaction de glissement
d Gleitreaktion
e fuerza de deslizamiento
n glijreactie

1102 slip
f glissement
d Schlupf; Gleitung
e desprendimiento; escape; pérdida
n slip

* **slipforms** → 1100

1103 slipway
f cale de construction; cale de lancement
d Schiffshelling
e vía de carena
n scheepshelling

1104 slope
f talus
d Böschung
e talud
n talud

1105 slope (angle)
f pente
d Gefälle; Neigung
e pendiente
n helling (hoek)

1106 slot
f fente
d Schlitz
e ranura; muesca; hendidura
n sleuf

1107 slow-setting cement
f ciment à prise lente
d langsam abbindender Zement
e cemento de fraguado lento
n langzaam bindend cement

1108 sluice
f écluse
d Schütze
e esclusa; conducto de evacuación
n sluis

1109 slump
f affaissement
d Setzmass
e asiento; aplastamiento; hundimiento
n zetmaat

1110 slump cone
f cône d'Abrams
d Setzbecher
e molde cónico para la prueba de asiento; cono de Abrams
n kegel van Abrams

1111 slump test
f essai d'affaissement
d Setzversuch

e prueba de asiento
n zetproef

1112 snow load(ing)
f charge de neige
d Schneebelastung
e carga de nieve
n sneeuwbelasting

1113 solid (not fluid)
f solide
d fest
e sólido
n vast

1114 solid (not hollow)
f plein; massif
d massiv; voll
e macizo
n massief

1115 solid-web girder; plain girder
f poutre à âme pleine
d Vollwandträger
e viga de alma llena
n vollewandligger

1116 sound *v*
f sonder
d sondieren
e sondar; sondear
n sonderen

1117 soundness
f solidité
d Güte
e solidez
n deugdelijkheid

1118 soundness; volume stability
f constance de volume
d Raumbeständigkeit
e inalterabilidad
n vormhoudendheid

1119 spacer
f espaceur; écarteur
d Abstandhalter
e separador
n afstandhouder

1120 spacer block
f cale
d Abstandhalter
e separador; taco; calzo
n dekkingsblokje

1121 spacer ring
f anneau espaceur
d ringförmiger Abstandhalter
e separador anular
n afstandsring

1122 spalling
f épaufrure
d Absplittern; Abplatzen
e descantillamiento
n afboeren

1123 span
f portée; travée
d Spannweite; Feld
e tramo; vano
n overspanning; veld

1124 spanning direction
f sens de la portée
d Tragrichtung
e dirección del vano
n draagrichting

1125 specific
f spécifique
d spezifisch
e específico
n specifiek

1126 specific gravity
f poids spécifique
d spezifisches Gewicht
e peso específico
n soortelijk gewicht

1127 specific heat
f chaleur spécifique
d spezifische Wärme
e calor específico
n soortelijke warmte

1128 specification and conditions (of contract)
f cahier des charges et stipulations (d'un contrat)
d Leistungsbeschreibung und Lastenheft (eines Vertrages
e especificaciones y condiciones (de un contrato)
n bestek en voorwaarden (van een contract)

1129 spherical shell
f coque sphérique
d Kugelschale
e lámina esférica
n bolschaal

1130 spherical stress
f contrainte sphérique
d Kugelspannung
e tensión esférica
n bolspanning

1131 spiral reinforcement
f armature spirale
d Spiralbewehrung
e armadura en espiral; zuncho; armadura helicoidal
n spiraalwapening

1132 spiral spring
f ressort spiral
d Spiralfeder
e resorte en espiral; muelle en espiral
n spiraalveer

1133 spirally reinforced column; column with helical binding
f colonne frettée
d umschnürte Säule
e soporte zunchado; soporte con armadura helicoidal
n omwikkelde kolom

1134 splice plate
f couvre-joint
d Verbindungslasche
e placa de empalme
n koppelplaat

1135 splicing loop
f boucle de couplage
d Verbindungsschlaufe
e lazo de empalme
n verbindingslas

1136 splicing sleeve
f manchon de couplage
d Verbindungsmuffe
e manguito de empalme
n verbindingsmof; koppelmof

1137 split *v*
f fendre
d spalten
e hender; rajar
n splijten

1138 splitting tensile strength
f résistance à la traction par fendage
d Spaltzugfestigkeit
e resistencia de hendimiento
n splijttrekvastheid

1139 spool
f bobine
d Spule
e carrete; devanadera
n klos

1140 spot weld
f soudure par points
d Punktschweissung
e punto de soldadura
n puntlas

1141 spread anchorage
f ancrage en éventail
d Spreizverankerung
e anclaje en abanico
n spreidverankering

1142 spring constant
f constante d'élasticité
d Federkonstante
e constante de un muelle
n veerconstante

1143 square bar
f barre de section carrée
d Vierkantstab
e barra cuadrada; cuadradillo
n vierkant staal

1144 stability
f stabilité
d Stabilität; Standsicherheit
e estabilidad
n stabiliteit

1145 stabilize *v*
f stabiliser
d stabilisieren
e estabilizar
n stabiliseren

1146 stack (up) *v*
f entasser
d aufschichten
e apilar; amontonar
n optassen

1147 stage
f étape; phase
d Bauabschnitt; Stadium
e etapa; fase; escalón
n etappe; fase

1148 standard
f norme
d Norm
e norma
n norm

1149 standard deviation
f écart quadratique moyen
d Standardabweichung
e desviación standard
n standaardafwijking

1150 standard strength
f résistance normale
d Normfestigkeit
e resistencia normal
n normsterkte

1151 standardization
f normalisation
d Normung
e normalización
n normalisatie

1152 standardize *v*
f normaliser
d normen; normalisieren
e normalizar; uniformar
n normaliseren

1153 state of stress
f état de contrainte
d Spannungszustand
e estado de tensiones
n spanningstoestand

1154 static load(ing)
f charge statique
d ruhende Belastung
e carga estática
n rustende belasting

1155 static moment
f moment statique
d statisches Moment
e momento estático
n statisch moment

1156 statically determinate
f isostatique
d statisch bestimmt
e isostático
n statisch bepaald

1157 statically indeterminate; hyperstatic
f hyperstatique
d statisch unbestimmt
e hiperestático
n hyperstatisch

1158 statics
f statique
d Statik
e estática
n statica

1159 statistics
f statistique
d Statistik
e estadística; datos estadísticos
n statistiek

1160 stay
f hauban
d Abspannung
e puntal; viento; tirante
n tui

1161 steam
f vapeur
d Dampf
e vapor
n stoom

1162 steam curing
f traitement à la vapeur; étuvage
d Dampferhärtung
e curado al vapor
n stoomverharding

1163 steel
f acier
d Stahl
e acero
n staal

1164 steel plate
f tôle d'acier (tôle forte)
d Stahlblech (Grobblech)
e plancha de acero
n staalplaat (dik)

1165 steel sheet
f tôle d'acier (tôle mince)
d Stahlblech (Feinblech)
e lámina de acero
n staalplaat (dun)

1166 steel wire
f fil d'acier
d Stahldraht
e alambre de acero
n staaldraad

1167 steel wire rope
f câble d'acier
d Stahldrahtseil
e cable de alambre de acero
n staaldraadkabel

1168 steelfixer
f ferrailleur
d Eisenflechter
e alambre de atado
n vlechter

1169 steelfixer's pliers
f pinces de ferraillage
d Flechtzange
e tenazas para el atado con alambre
n vlechttang

1170 stiff concrete
f béton sec
d erdfeuchter Beton
e hormigón endurecido; hormigón seco
n aardvochtig beton

1171 stiffen *v*
f raidir
d versteifen
e endurecer; rigidizar; reforzar
n verstijven

1172 stiffness
f raideur
d Steifheit
e rigidez
n stijfheid

1173 stirrup
f étrier
d Bügel
e cerco; estribo
n beugel

1174 stirrup spacing
f espacement des étriers
d Bügelabstand
e distancia entre cercos
n beugelafstand

1175 stone chippings
f concassé
d Splitt
e grava; cascajo
n steenslag

1176 storage
f stockage
d Lagerung; Lagerhaltung
e almacenamiento
n opslag

1177 storey
f étage
d Stockwerk
e piso; planta
n verdieping

1178 strain
f déformation
d Dehnung
e deformación unitaria
n rek; vormverandering

1179 strain gauge
f jauge de déformation
d Dehnungsmesser
e elongámetro de resistencia eléctrica
n rekmeter

1180 strand
f toron
d Litze
e cordón
n streng

1181 strength
f résistance
d Festigkeit
e resistencia
n sterkte; vastheid

1182 strength of materials
f résistance des matériaux
d Materialfestigkeit
e resistencia de materiales
n materiaalsterkte

1183 strength under sustained load
f résistance sous charge de longue durée
d Dauerstandfestigkeit
e resistencia bajo carga mantenida
n langeduursterkte

1184 stress
f contrainte; tension
d Spannung
e tensión
n spanning

1185 stress circle (Mohr)
f cercle de Mohr
d Spannungskreis (Mohr)
e círculo de tensiones (Mohr)
n spanningscirkel (Mohr)

1186 stress concentration
f concentration de contrainte
d Spannungskonzentration
e concentración de tensiones
n spanningsconcentratie

1187 stress corrosion
f corrosion sous tension
d Spannungskorrosion
e corrosión bajo tensión
n spanningscorrosie

1188 stress decrease
f diminution de la contrainte
d Spannungsverminderung
e disminución de tensiones
n spanningsdaling

1189 stress distribution
f distribution des contraintes
d Spannungsverteilung
e distribución de tensiones
n spanningsverdeling

1190 stress gradient
f gradient de contrainte
d Spannungsgefälle
e gradiente de tensiones
n spanningsgradiënt

1191 stress jump
f saut de contrainte
d Spannungssprung
e salto de tensión
n spanningssprong

1192 stress loss
f perte de contrainte
d Spannungsverlust
e pérdida de tensión
n spanningsverlies

1193 stress region
f zone de contrainte
d Spannungsbereich
e zona de tensión
n spanningsgebied

1194 stress-strain diagram
f diagramme contrainte-déformation
d Spannungs-Dehnungs-Diagramm
e diagrama tensión-deformación
n spanning-rekdiagram

1195 stress trajectories
f trajectoires des contraintes
d Spannungstrajektorien
e trayectorias de tensiones
n spanningstrajectoriën

1196 stress variation
f variation des contraintes
d Spannungsschwankung
e variación de tensiones
n spanningsvariatie

1197 stress wave
f onde de contrainte
d Spannungswelle
e onda de tensión
n spanningsgolf

1198 stressed-ribbon bridge
f pont-ruban précontraint
d Spannbandbrücke
e puente tipo "banda pretensada"
n spanbandbrug

1199 stressing abutment
f massif d'ancrage (prétension)
d Ankerbock
e estribo de tesado
n verankeringslichaam

1200 stressing equipment
f équipement de mise en tension
d Spannvorrichtung
e equipo de tesado
n spanapparatuur

1201 stressing head
f tête de mise en tension
d Spannkopf
e cabeza de tesado
n spankop

1202 stressing jack
f vérin de mise en tension
d Spannpresse
e gato de tesado
n spanvijzel

1203 stressing operative
f technicien pour la mise en tension
d Spannmonteur
e operario que realiza el tesado
n spanmonteur

1204 stressing scheme
f schéma de mise en tension
d Spannplan
e esquema de tesado; plan de tesado
n spanschema

1205 stretch *v*
f étendre
d dehnen
e estirar
n rekken

1206 strike *v*; **strip** *v* (formwork)
f décoffrer
d ausschalen
e desmoldar; descimbrar
n ontkisten; lossen

1207 strike off *v*
f régler; araser
d abgleichen
e enrasar; nivelar
n afrijen

* **strip** *v* (formwork) → 1206

1208 structural member
f élément de structure
d Bauteil
e elemento estructural
n constructiedeel

1209 structural strength reserve
f réserve de résistance
d Gestaltfestigkeit
e reserva de resistencia de una estructura
n incasseringsvermogen

1210 structure
f structure; ouvrage
d Bauwerk
e estructura
n bouwwerk; constructie

* **strut** → 0916

1211 subsoil
f sous-sol
d Untergrund
e subsuelo
n ondergrond

1212 substructure
f infrastructure
d Unterbau
e infraestructura
n onderbouw

1213 subterranean
f souterrain
d unterirdisch
e subterráneo
n ondergronds

1214 superimpose *v*
f superposer
d überlagern
e superponer
n superponeren

* **superimposed load(ing)** → 0723

1215 supersulphated cement
f ciment sursulfaté
d Sulfathüttenzement
e cemento sobresulfatado
n gesulfateerd cement

1216 supervision
f contrôle
d Aufsicht

e inspección; supervisión
n toezicht

1217 supervisor
f contrôleur
d Aufseher
e inspector; supervisor
n opzichter

1218 support
f appui
d Auflager; Auflagerung; Stützung
e apoyo; soporte; puntal
n ondersteuning; steunpunt

1219 support moment
f moment d'appui
d Stützmoment
e momento en el apoyo
n steunpuntsmoment

1220 support reaction
f réaction d'appui
d Auflagerreaktion
e reacción en el apoyo
n oplegreactie

1221 surface
f surface
d Oberfläche
e superficie
n oppervlak

1222 surface reinforcement
f armature de peau
d Oberflächenbewehrung
e armadura superficial
n huidwapening

1223 surface tension
f tension superficielle
d Oberflächenspannung
e tensión superficial
n oppervlaktespanning

1224 surface treatment
f traitement de surface
d Oberflächenbehandlung
e tratamiento superficial
n oppervlaktebehandeling

1225 surface vibrator
f vibrateur de surface
d Oberflächenrüttler
e vibrador superficial
n oppervlaktetriller

• **suspended beam** → 0431

1226 suspended roof
f toiture suspendue
d Hängedach
e cubierta colgante
n hangdak

1227 suspension bridge
f pont suspendu
d Hängebrücke
e puente colgante
n hangbrug

1228 swell *v*
f gonfler
d schwellen; quellen
e inflar; dilatar; entumecer
n zwellen

1229 synthetic resin
f résine synthétique
d Kunstharz
e resina sintética
n kunsthars

T

1230 tack welding
f pointage
d Haftschweissung
e soldadura provisional
n hechtlassen

1231 tamp *v*
f damer
d stampfen
e golpear; apisonar
n stampen; porren

1232 tamped concrete
f béton damé
d Stampfbeton
e hormigón apisonado
n stampbeton

1233 tangent
f tangent
d Tangente
e tangente
n raaklijn

1234 tangent point
f point de contact
d Tangentenberührungspunkt; Berührungspunkt
e punto de tangencia
n raakpunt

1235 tangential
f tangentiel
d tangential
e tangencial
n tangentiaal

1236 tank
f réservoir
d Behälter
e tanque; depósito
n reservoir

1237 taper *v*
f effiler
d verjüngen
e ahusar; afilar; acabar en punta
n verjongen

1238 tar epoxy
f epoxy au goudron
d teerepoxy
e alquitrán epoxy
n teerepoxy

1239 temper *v*
f tremper
d tempern
e templar
n ontlaten

1240 temperature gradient
f gradient de température
d Temperaturgefälle
e gradiente de temperatura
n temperatuurgradiënt

1241 temperature shrinkage
f retrait thermique
d Temperaturschwindung
e retracción térmica
n temperatuurkrimp

1242 temperature stress
f contrainte thermique
d temperaturspannung
e tensión térmica
n temperatuurspanning

* **tender** → 0116

1243 tendon (prestressing)
f câble (de précontrainte)
d Spannglied
e tendón (de pretensado)
n spanelement

1244 tendonless prestressing
f précontrainte sans câbles
d kabellose Vorspannung
e pretensado sin armadura
n kabelloze voorspanning

1245 tensile force
f force de traction
d Zugkraft
e fuerza de tracción
n trekkracht

1246 tensile strain
f déformation de traction
d Dehnung
e deformación unitaria por tracción
n rek

1247 tensile strength
f résistance à la traction
d Zugfestigkeit
e resistencia a tracción
n treksterkte; trekvastheid

1248 tensile stress
f contrainte de traction
d Zugspannung
e tensión de tracción
n trekspanning

1249 tensile test
f essai à la traction
d Zugversuch
e ensayo de tracción
n trekproef

1250 tensile zone
f zone tendue
d Zugzone
e zona de tracción
n trekzone

1251 tension *v*
f mettre en traction
d spannen
e tesar
n spannen

1252 tension
f traction
d Zug
e tensión; tracción
n trek

1253 tension pile
f pieu tendu
d Zugpfahl
e pilote de tracción
n trekpaal

1254 tension reinforcement
f armature de traction
d Zugbewehrung
e armadura de tracción
n trekwapening

1255 tensioning equipment
f équipement de mise en tension
d Spannvorrichtung
e equipo de tesado
n spanapparatuur

1256 tensioning jack
f vérin de mise en tension
d Spannpresse
e gato de tesado
n spanvijzel

1257 test
f essai; expérience
d Versuch
e ensayo; prueba
n proef

1258 test bar
f éprouvette (barre)
d Versuchsstab; Probestab
e barra de ensayo
n proefstaaf

1259 test beam
f poutre d'essai
d Probebalken
e viga de ensayo
n proefbalkje

1260 test cube
f cube d'essai
d Probewürfel
e probeta cúbica
n proefkubus

1261 test cylinder
f cylindre d'essai
d Probezylinder
e probeta cilíndrica
n proefcilinder

1262 test load(ing)
f charge d'essai
d Prüflast
e carga de ensayo
n proefbelasting

1263 test piece; test specimen
f éprouvette
d Probekörper; Probestück
e probeta de ensayo
n proefstuk

* **test specimen** → 1263

1264 testing
f essai
d Prüfung
e ensayo; prueba
n onderzoek; beproeving

1265 theoretical mechanics
f mécanique théorique
d theoretische Mechanik
e mecánica teórica
n theoretische mechanica

1266 thickness
f épaisseur
d Dicke; Stärke
e espesor; grueso
n dikte

1267 thixotropic
f thixotrope
d thixotrop
e tixotropía
n thixotroop

1268 three-hinged frame; three-pinned frame
f portique à trois articulations
d Dreigelenkrahmen
e pórtico triarticulado; estructura triarticulada
n driescharnierspant

* **three-pinned frame** → 1268

1269 thrust
f poussée
d Horizontalschub; Schub
e empuje; presión
n spatkracht

1270 thrust-relieving arch
f voûte de décharge
d Druckbogen
e arco de apoyo de una bóveda
n drukboog

1271 tie (member)
f tirant
d Zugband; Zugglied
e atadura; tirante
n trekband

1272 tie (of reinforcement)
f ligaturer (ferraillage)
d binden (der Stahleinlagen); rödeln (der Stahleinlagen)
e tirar (de armadura); cercar (en soportes)
n binden (van wapening)

1273 tie; sleeper (railroad)
f traverse (chemin de fer)
d Schienenschwelle (Eisenbahn)
e traviesa (de ferrocarril)
n dwarsligger (spoorwegen)

1274 tie-rod
f tirant
d Zugstange
e tirante
n trekstang

1275 tie-wire
f fil de ligature
d Bindedraht; Rödeldraht
e alambre de atadura; viento
n binddraad

1276 tilting drum (of mixer)
f tambour basculant (mélangeur à béton)
d Kipptrommel (des Mischers)
e cuba basculante (de amasadora); tambor basculante (de amasadora)
n kantelbare mengtrommel

* **timber** → 1402

1277 tolerance
f tolérance
d Toleranz
e tolerancia
n tolerantie

1278 tools
f outillage
d Werkzeug; Geräte
e herramientas
n gereedschap

1279 tooth
f dent
d Zahn
e diente
n tand

1280 top bar
f barre supérieure
d oberer Stab
e barra superior
n bovenstaaf

1281 top fibre
f fibre supérieure
d obere Faser
e fibra superior
n bovenste vezel

1282 top flange
f membrure supérieure
d Obergurt; Oberflansch
e ala superior
n bovenflens

1283 top mesh
f treillis supérieur
d oberes Bewehrungsnetz
e malla superior
n bovennet

1284 topping
f béton de plancher à hourdis
d Aufbeton
e acabado; capa de rodadura
n druklaag

1285 torque wrench
f clé dynamométrique
d Drehmomentschlüssel
e llava de torsión; llava dinamométrica
n momentsleutel

1286 torsional moment
f moment de torsion
d Torsionsmoment
e momento torsor; momento de torsión
n wringmoment

1287 torsional stress
f contrainte de torsion
d Torsionsspannung
e tensión de torsión
n wringspanning

1288 traffic load
f charge roulante
d Verkehrsbelastung; Verkehrslast
e carga de tráfico; sobrecarga
n verkeersbelasting

1289 transmission zone
f zone de transmission
d Einleitungszone
e zona de transmisión
n overdrachtszone

1290 transverse
f transversal
d quer
e transversal
n dwars

* **transverse contraction** → 0695

1291 transverse crack
f fissure transversale
d Querriss
e fisura transversal
n dwarsscheur

1292 transverse member
f entretoise; traverse
d Riegel
e elemento transversal
n dwarsstaaf

1293 transverse prestress(ing)
f précontrainte transversale
d Quervorspannung
e pretensado transversal
n dwarsvoorspanning

1294 transverse reinforcement
f armature transversale
d Querbewehrung
e armadura transversal
n dwarswapening

1295 trass
f trass
d Trass
e trass
n tras

1296 trass Portland cement
f ciment portland au trass
d Trass-Portlandzement
e cemento portland de trass; cemento de trass
n tras-portlandcement

1297 travel (of a jack)
f course (d'un vérin)
d Hub (einer Presse)
e recorrido (de un gato)
n slag (van een vijzel)

1298 tremie
f trémie
d Betoniertrichter; Trichter

e tolva; embudo
n stortkoker

1299 tremie method (under-water concreting)
f bétonnage à la trémie (sous l'eau)
d Kontraktorverfahren (Unterwasser-Betonieren)
e método "trémie" (hormigonado bajo el agua)
n contractormethode

1300 trestle frame
f chevalet; palée
d Joch
e palanca
n juk

1301 triaxial prestress
f précontrainte triaxiale
d dreiachsige Vorspannung
e pretensión triaxil
n triaxiaalvoorspanning

1302 triaxial stress
f contrainte triaxiale
d dreiachsige Spannung
e tensión triaxil
n triaxiaalspanning

1303 trowel
f truelle
d Kelle
e llana; palustre
n troffel

1304 truck-mixer
f camion malaxeur
d Fahrmischer
e camión mezclador; camión agitador
n truckmixer

* **truss** → 0696

* **truss** → 0697

1305 tunnel
f tunnel
d Tunnel
e túnel
n tunnel

1306 turnbuckle
f manchon de serrage
d Spannschloss
e manguito tensor
n wartel

1307 twist *v*
f tordre
d verdrehen
e torcer; retorcer
n wringen; torderen

U

1308 U-hook
f crochet considère
d halbkreisförmiger Haken
e gancho en U
n ronde haak

1309 ultimate compressive stress
f résistance à la compression
d Druckgrenze
e compresión de rotura
n breukstuik

1310 ultimate load analysis
f calcul à la rupture
d Traglastverfahren
e cálculo en rotura
n bezwijkanalyse

1311 ultimate moment
f moment de rupture
d Bruchmoment
e momento de rotura
n breukmoment

1312 ultimate strength
f résistance à la rupture
d Bruchfestigkeit
e resistencia a rotura
n breuksterkte

1313 ultimate stress
f contrainte de rupture
d Bruchspannung
e tensión de rotura
n breukspanning

1314 ultimate tensile strain; elongation at fracture
f allongement à la rupture
d Bruchdehnung
e deformación unitaria de rotura por tracción
n breukrek

1315 unbonded tendon
f câble non adhérent
d Spannglied ohne Verbund
e tendón sin adherencia; tendón no adherente
n niet geïnjecteerde kabel

1316 uncoil *v*
f dérouler
d abwickeln
e desenrollar
n afwikkelen

1317 underground
f souterrain
d unterirdisch
e subterráneo
n ondergronds

1318 underwater concrete
f béton sous l'eau
d Unterwasserbeton
e hormigón sumergido
n onderwaterbeton

1319 undo *v*
f défaire
d lösen
e deshacer; desprender
n losmaken

1320 uniform load
f charge uniforme
d Gleichlast
e carga uniforme
n gelijkmatige belasting

1321 uniformly distributed load(ing)
f charge uniformément répartie
d gleichmässig verteilte (Be)last(ung)
e carga uniformemente distribuída
n gelijkmatig verdeelde belasting

1322 unit
f unité
d Einheit
e unidad
n eenheid

* **unreinforced** → 0860

1323 unscrew *v*
f dévisser
d losschrauben; abschrauben
e destornillar
n losdraaien

1324 unskilled labourer; unskilled workman
f main-d'oeuvre non spécialisée

d ungelernter Arbeiter; Hilfsarbeiter
e obrero inexperto; obrero no especializado
n ongeschoolde arbeider

* **unskilled workman** → 1324

1325 unstressed
f sans contrainte
d spannungslos
e no teso; sin tensión
n spanningloos

1326 untensioned reinforcement
f armature précontrainte
d schlaffe Bewehrung
e armadura sin tensión
n niet-voorgespannen wapening

1327 unwind *v*
f dérouler
d abwickeln
e desenrollar
n afwikkelen

1328 upset *v*
f refouler
d stauchen
e recalcar; engrosar
n opstuiken

1329 upsetting limit
f limite de refoulement
d Stauchgrenze
e límite de relajamiento
n stuikgrens

V

1330 vacuum concrete
f béton traité sous vide
d Vakuumbeton
e hormigón al vacío
n vacuumbeton

1331 variable
f variable
d variabel
e variable
n variabel

1332 vault
f voûte
d Gewölbe
e bóveda
n gewelf

1333 vector
f vecteur
d Vektor
e vector
n vector

* **vent** *v* → 0124

1334 vent (hole)
f trou d'évent; évent
d Entlüftungsöffnung
e respiradero; tubo de purga
n ontluchtingsopening

1335 vent pipe
f tube d'évent
d Entlüftungsrohr
e tubo de ventilación
n ontluchtingspijpje

1336 vertical
f vertical
d vertikal; senkrecht
e vertical
n loodrecht

1337 vertical section
f section verticale
d Vertikalschnitt
e sección vertical
n verticale doorsnede

1338 viaduct
f viaduc
d Viadukt
e viaducto
n viaduct

1339 vibrate *v*
f vibrer
d rütteln
e vibrar; compactar por vibración
n trillen

1340 vibrating and finishing machine
f vibro-finisseur
d Rüttelfertiger
e máquina vibrante; vibrador
n trilmachine (wegenbouw)

1341 vibrating beam
f poutre vibrante
d Rüttelbohle
e viga vibrante
n trilbalk

1342 vibrating table
f table vibrante
d Rütteltisch
e mesa vibrante; mesa de sacudidas
n triltafel

1343 vibration
f vibration
d Schwingung
e vibración
n trilling

1344 vibrator
f vibrateur
d Rüttler
e vibrador
n triller

1345 vibratory screen
f crible vibrant
d Schwingsieb
e tamíz vibrante
n trilzeef

1346 Vierendeel-girder
f poutre Vierendeel
d Vierendeelträger
e viga Vierendeel
n Vierendeelligger

1347 viscosity
f viscosité
d Viskosität
e viscosidad
n viscositeit

1348 voids; pores
f pores; vides
d Poren
e cavidades; huecos; poros
n poriën; luchtporiën

1349 voids content; pores content
f teneur en vides
d Porengehalt
e contenido de huecos
n poriëngehalte

1350 voids volume; pore volume; pore space
f volume des vides
d Porenvolumen; Porenraum
e volúmen de huecos
n poriënvolume; poriënruimte

1351 volume
f volume
d Volumen
e volúmen
n volume

1352 volume-batching
f dosage volumétrique
d Volumendosierung; Raumdosierung
e dosificación; volumétrica
n dosering naar maatdelen

* **volume stability** → 1118

W

1353 waffle-slab floor
f plancher en dalle nervurée
d Kassettendecke
e forjado reticulado; forjado de casetones
n cassettevloer

1354 waling
f moise
d Holm
e larguero
n gording (van beschoeiing)

1355 wall
f mur
d Wand
e muro; pared; tabique
n wand

1356 warped
f gauchi
d windschief
e alabeada
n scheluw

1357 warping
f gauchissement
d Verwerfung
e alabeo
n kromtrekking

1358 washer
f rondelle
d Beilagscheibe; Unterlegscheibe
e arandela
n vulring; onderlegplaatje

1359 water-cement ratio
f facteur eau-ciment
d Wasser-Zement-Wert
e relación agua-cemento
n watercementfactor

1360 water content
f teneur en eau
d Wassergehalt
e contenido de agua
n watergehalte

* **water segregation** → 0125

1361 water-stop
f bande d'arrêt d'eau
d Fugenband
e tapajunta
n voegstrip

1362 water supply
f approvisionnement en eau
d Wasserversorgung
e suministro de agua; abastecimiento de agua
n watervoorziening

1363 water tower
f château d'eau
d Wasserturm
e depósito elevado de agua
n watertoren

* **waterproof** → 1364

1364 watertight; waterproof
f étanche
d wasserdicht
e estanco; impermeable
n waterdicht

1365 watertight concrete
f béton étanche
d wasserundurchlässiger Beton
e hormigón impermeable
n waterdicht beton

1366 watertightness
f étanchéité
d Wasserundurchlässigkeit
e estanquidad
n waterdichtheid

1367 wear
f usure
d Verschleiss; Abnutzung
e desgaste; uso
n afslijting; slijtage

1368 weather code
f code météorologique
d Wetterkode
e previsión meteorológica; parte meteorológico
n weercode

1369 weather-resistant
f résistant aux intempéries

d wetterbeständig
e resistente a la intemperie
n weerbestendig

1370 web (of a beam)
f âme (d'une poutre)
d Steg (eines Trägers)
e alma (de una viga)
n lijf (van een balk)

1371 wedge
f coin; clavette
d Keil
e cuña
n wig

1372 wedge anchorage
f ancrage par clavettes
d Keilverankerung
e anclaje de cuña
n wigverankering

1373 wedging jack
f vérin à clavettes
d Kolben zur Festsetzung der Keile
e gato de acuñamiento
n blokkeervijzel

1374 weigh *v*
f peser
d wiegen
e pesar
n wegen

1375 weigh-batching
f dosage en poids
d Gewichtsdosierung; Wiegedosierung
e dosificación
n dosering naar gewichtsdelen

1376 weigh-hopper
f trémie de pesage
d Wiegebehälter
e tolva dosificadora en peso
n weegbak

1377 weighing apparatus
f balance de pesée
d Wiegevorrichtung
e aparato de pesar
n weegapparaat

1378 weight
f poids
d Gewicht
e peso
n gewicht

1379 weir
f barrage
d Wehr
e vertedero; azud
n stuw

1380 weld *v*
f souder
d schweissen
e soldar
n lassen

1381 weld bead
f cordon de soudure
d Schweissraupe
e cordón de soldadura
n lasrups

1382 weld length
f longueur de soudure
d Schweisslänge
e longitud de soldadura
n laslengte

1383 weldable
f soudable
d schweissbar
e soldable
n lasbaar

1384 welding equipment
f matériel de soudage
d Schweissanlage
e equipo de soldadura
n lasapparatuur

1385 white cement
f ciment blanc
d Weisszement
e cemento blanco
n wit cement

1386 white sand
f sable blanc
d Weissand
e arena argentífera
n zilverzand

1387 width; breadth
f largeur
d Breite
e anchura
n breedte

1388 wind
f vent
d Wind
e viento
n wind

1389 wind bracing
f contreventement
d Windverband
e contraviento; arriostramiento
n windverband

1390 wind load(ing)
f charge de vent
d Windbelastung
e carga de viento
n windbelasting

1391 wind pressure
f pression du vent
d Winddruck
e presión del viento
n winddruk

1392 winding machine
f machine à enrouler
d Wickelmaschine
e bobinadora; devanadora
n wikkelmachine

1393 wing nut
f écrou à oreilles
d Flügelmutter
e tuerca de orejas; mariposa
n vleugelmoer

1394 wire
f fil
d Draht
e alambre
n draad; ijzerdraad

1395 wire anchorage
f ancrage de fil
d Drahtverankerung
e anclaje del alambre
n draadverankering

1396 wire clamp
f serre-fil
d Drahtklemme
e grupo de alambre
n draadklem

1397 wire coil
f couronne de fil
d Drahtring
e rollo de alambre; carrete de alambre
n draadring

1398 wire reel
f bobine de fil
d Drahthaspel
e espira de alambre
n draadhaspel

1399 wire sieve
f tamis à fil
d Drahtsieb
e cedazo de malla de alambre
n draadzeef

1400 wire winding machine (for tanks)
f machine d'enroulement (précontrainte de réservoir)
d Drahtwickelmaschine (für Behälter)
e máquina devanadora de alambres (para el pretensado de depósitos circulares)
n draadwikkelmachine (voor reservoirs)

1401 wobble effect (tendon friction)
f frottement due aux petites ondulations aléatoires des gaines
d Reibung infolge zufälliger Welligkeit des Spannkanals
e efecto ondulante (rozamiento parásito, en las partes rectas del trazado de las armaduras de pretensado)
n wobble-effect (wrijving door kleine afwijkingen van het kanaaltracé)

1402 wood; timber
f bois
d Holz
e madera
n hout

1403 wooden float
f taloche en bois
d Reibbrett; Holzspachtel
e llana de albañil
n plekspaan

1404 workable (weather)
f ouvrable (temps)
d brauchbar (Arbeitswetter)
e laborable (tiempo)
n werkbaar (weer)

1405 workability
f maniabilité; ouvrabilité
d Verarbeitbarkeit
e trabajabilidad
n verwerkbaarheid

1406 working drawing
f dessin d'exécution
d Arbeitszeichnung
e plano de trabajo
n werktekening

1407 working stress
f tension de travail
d Gebrauchsspannung
e tensión de trabajo
n werkspanning

1408 workman; labourer
f ouvrier
d Arbeiter
e obrero; operario
n werkman; arbeider

1409 workshop
f atelier
d Werkstatt
e taller
n werkplaats

1410 worm conveyor; screw conveyor
f transporteur à vis; vis transporteuse
d Förderschnecke; Schneckenförderer
e tornillo transportador
n transportschroef

1411 wrap *v*
f envelopper
d umwickeln
e ceñir; rodear; enrollar; bobinar
n omwikkelen

X

1412 X-ray examination
f contrôle aux rayons-X
d Röntgenuntersuchung
e radiografía; examen por rayos X
n röntgenologisch onderzoek

Y

1413 yield line
f ligne de rupture
d Bruchlinie
e línea de cedencia
n vloeilijn

1414 yield point
f limite élastique
d Streckgrenze; Fliessgrenze
e límite de cedencia
n rekgrens; vloeigrens

1415 yield range
f domaine d'écoulement
d Fliessbereich
e zona de cedencia
n vloeigebied

1416 yielding (of steel)
f écoulement (de l'acier)
d Fliessen (von Stahl)
e cedencia (del acero)
n vloeien (van staal)

FRANÇAIS

DEUTSCH

ESPAÑOL

NEDERLANDS

РУССКОЕ ЧИСЛЕННОЕ ОГЛАВЛЕНИЕ

RUSSIAN NUMERICAL INDEX

1 сопротивление истиранию
2 абсорбция
3 водопоглощение
4 опора, устой
5 ускорять
6 кислотность
7 кислотостойкий
8 акустические испытания
9 прибавлять
10 изоляционная лента
11 регулировать
12 установочный винт
13 добавка
14 возраст
15 агент, вещество
16 заполнитель (бетона)
17 агрессивный
18 мешалка
19 воздух; вентилировать
20 воздушный пузырек
21 помещение с кондиционированным воздухом
22 содержание воздуха
23 воздухововлекающая добавка
24 воздухомер
25 воздухонепроницаемый бетон
26 пустоты
27 щелочной
28 сплав
29 знакопеременная нагрузка
30 знакопеременное напряжение
31 глинозёмистый цемент
32 амплитуда
33 анкер
34 плита анкерной опоры
35 анкерный конус
36 анкерная коробка
37 анкерная втулка
38 анкерный вкладыш
39 анкерная гайка
40 анкерная плита
41 анкерный клин
42 анкеровка
43 длина зоны анкеровки
44 зона анкеровки
45 угол (в градусах)
46 угол трения
47 угол скручивания
48 угол вращения
49 анизотропный
50 отжигать, отпускать
51 отожженная проволока
52 объемная масса
53 береговой пролет
54 акведук
55 арка
56 арочный мост
57 архитектор
58 плечо (пары сил)
59 искусственное старение
60 искусственная пена
61 монтировать, собирать
62 асимптота
63 прикреплять
64 автоклав
65 вспомогательная конструкция
66 нагрузка от снежной лавины
67 среднее число
68 среднее напряжение
69 получение контракта
70 осевой
71 осевая нагрузка
72 осевое напряжение
73 ось (графика)
74 нагрузка на ось
75 равновесие, весы
76 уравновешивающая балка, коромысло весов
77 балконная плита
78 шаровая опора
79 шаровое или сферическое соединение
80 шаровая мельница
81 стержень
82 сетка из стержней
83 поперечное сечение стержня
84 просвет между стержнями
85 заграждение, дамба
86 бочарный свод
87 подошва (фундамента)
88 основание, подвал
89 основа (расчета)
90 дозировка
91 пролет (в здании)
92 балка (брус)
93 решетка из балок
94 поперечное сечение балки
95 опора, подшипник
96 опорная подушка (пролетного строения моста)
97 длина опирания
98 опорное давление
99 реакция опоры
100 ленточный транспортер
101 отгибать вниз

102 оттягивать вверх (канатную арматуру)
103 арматурный цех
104 гибочный станок (для стержней)
105 гибочный пресс (для стержней)
106 изгибающий момент
107 эпюра изгибающих моментов
108 режим изгиба (стержей)
109 испытание на изгиб
110 отогнутый
111 отогнутый вниз стержень арматуры
112 отогнутый вверх стержень арматуры
113 отогнутая вверх арматура
114 скошенный (край)
115 двухосное напряжение
116 заявка (на контракт)
117 счет, инвентарный список
118 сцеплять с арматурой
119 вяжущее, связующее вещество
120 скрепляющая арматура
121 битум
122 доменный, шлаковый цемент
123 доменный шлак
124 сочиться (о жидкости или газе)
125 выступание (отделение воды)
126 мелкий заполнитель, посыпка
127 болт
128 напряжение сцепления
129 вяжущее, скрепляющее вещество
130 нижний стержень
131 нижнее волокно
132 нижняя сетка
133 граничные условия
134 рамная конструкция коробчатого типа
135 балка коробчатого сечения
136 связь
137 кронштейн, вут
138 мост
139 мостовой настил
140 балка или ферма моста
141 мостовой бык (устой)
142 ломаный, лробленый
143 продольный изгиб (колонны)
144 профильный изгиб (плит)
145 коэффициент продольного изгиба
146 прочность при продольном изгибе
147 напряжение при продольном изгибе
148 здание, строительство
149 строительная администрация
150 владелец (здания) смотритель (здания)
151 строительные нормы
152 строительный надзор
153 бимсовая сталь
154 цемент насыпью
155 объемная масса
156 отжигать (проволоку)
157 обтеска бучардой
158 соединение встык
159 стыковая накладка
160 анкеровка высаженной головкой
161 контрфорс
162 канат, кабель, пучок
163 зажим троса, каната
164 каналообразователь
165 зажим каната
166 обшивка пучка
167 вантовый мост
168 кадмиевое покрытие, кадмирование
169 кессон
170 вычислять
171 калибровать
172 калибровочная величина
173 выпуклость, выгиб
174 консоль
175 консольный мост
176 консольная балка
177 выступать в виде консоли
178 выступаюшая в виде консоли конструкция (мостов)
179 головка пучка
180 ёмкость, мощность
181 капиллярный
182 картон
183 проезжая часть
184 обетонировать
185 чугун
186 стальное литье
187 полость, раковина
188 ячеистый бетон
189 цемент
190 химия цемента
191 цементный клинкер
192 содержание цемента
193 цементный гель
194 цемент насыпью (без упаковки)
195 цементный раствор
196 цементное тесто
197 банка, силос или бункер для цемента
198 цементная пленка
199 жидкое цементное тесто, цементный шлам
200 центрировать
201 средняя линия, ось (на графике)

202 центр тяжести (твердого тела)
203 центр двления
204 расстояние между центрами от оси до оси
205 бетон, отлитый центрифугированием
206 центрифугировать
207 центрирующий штифт
208 ось, проходящая через центр тяжести
209 сертификат (оценки)
210 характеристика
211 номинальный диаметр
212 нормативная прочность
213 контрольное испытание
214 химическое испытание
215 химически стойкий
216 главный инженер
217 крошка, мелкий щебень
218 резать, рубить
219 откалывать
220 литой бетон
221 окружность, периметр
222 периферическое предварительное напряжение
223 зажим
224 глина
225 внутренний диаметр, диаметр в свету
226 расстояние в свету
227 шуп
228 крупный заполнитель
229 крупный гравий
230 покрывать цементным раствором
231 коэффициент
232 коэффициент расширения
233 коэффициент трения
234 крэффициент изменчивости
235 цветной бетон
236 холоднотянутая стадь
237 холоднотянутая проволока
238 холоднообработанная сталь
239 сотрудничать
240 разрушаться
241 разрушающая нагрузка
242 колонна
243 ось колонны
244 база колонны
245 основание колонны
246 капитель колонны
247 полоса безбалочного перекрытия, перекрывающая колонну
248 колонна со связями (или затяжками)
249 товарное, торговое (рыночное) качество
250 уплотнять
251 показатель уплотнения
252 дополнительный
253 компонент
254 составлять
255 составная колонна
256 сжимать
257 сжатый воздух
258 сжатие, уплотнение
259 сжатый пояс
260 арматура, работающая на сжатие
261 машина для испытания на сжатие
262 сжимающая сила
263 прочность на сжатие
264 сжимаюшее напряжение
265 сжатая зона
266 компрессор
267 компьютер
268 сосредоточенная нагрузка
269 сходящийся пучок, кабель
270 бетон
271 Ассоциация по бетону
272 бетонный желоб
273 бетон, уплотненный встряхиванием
274 состав бетона
275 защитный слой бетона
276 бетоновод
277 машина для обработки поверхности бетона
278 бетонный шарнир
279 бетономешалка
280 бетонная смесь
281 покрывать бетонной смесью
282 бетонное дорожное покрытие
283 укладка бетонной смеси
284 бетонный завод
285 фанера для опалубки
286 проба бетона
287 бетононасос
288 бетонная распорка (прокладка, блок)
289 технология бетона
290 молоток для испытания бетона
291 бункер для бетонирования
292 ковш для бетонирования
293 условия сбыта
294 конус
295 соответствие (проектным) расчетным размерам
296 коническая оболочка
297 соединительная гайка

298 соединительная арматура
299 коноидальная оболочка
300 консистенция
301 рабочий шов
302 контакт, сцепление
303 контактный метод
304 контактная новерхность
305 вода затворения
306 загрязнять
307 загрязнение
308 содержание
309 содержимое
310 непрерывный (пучок)
311 момент неразрезности (на промежуточной опоре)
312 непрерывный, неразрезной
313 неразрезная балка
314 непрерывный гранулометрический состав
315 неразрезная плита
316 сварка непрерывным швом
317 подряд, логовор
318 цена по контракту
319 фирма-подрядчик
320 подрядчик
321 сужение, сжатие
322 выпуклый
323 охлаждать, остывать
324 способствовать, содействовать
325 координировать
326 выступ, выпуск, кронштейн
327 сердечник, стержень, ядро
328 радиус сердечника
329 угловой стержень
330 разъедать, ржаветь
331 коррозия
332 коррозийно-устойчивый
333 противотоковый смеситель
334 пара, пара сил
335 соединение, стяжка
336 покрывать
337 посыпать песком
338 трещина
339 форма трещин
340 расстояние между трещинами
341 ширина трещин
342 образование трещин
343 момент трещинообразования
344 кран
345 ползучесть
346 деформация ползучести
347 коэффициент ползучести
348 градиент ползучести
349 модуль ползучести
350 гофрировать
351 гофрированная проволока
352 критерий
353 поперечная балка
354 поперечная балка, ригель
355 поперечный элемент
356 поперечное сечение
357 крестообразное армирование
358 щебень
359 дробленый гравий
360 щебень, дробленый камень
361 дробление
362 куб
363 кубиковая прочность
364 бордюрный камень
365 выдерживать (бетон)
366 состав, наносимый для лучшего твердения бетона
367 кривизна
368 градиент кривизны
369 кривая
370 криволинейный
371 криволинейный кабель, пучок
372 резать
373 режим резки
374 циклическое нагружение
375 цилиндр
376 цилиндрическая прочность
377 цилиндрический
378 цилиндрическая оболочка
379 дамба
380 повреждение, дефект
381 глухой, концевой анкер
382 постоянная нагрузка
383 собственный вес
384 балка-стенка
385 прогибать, отклонять
386 оттягивать напрягаемую арматуру
387 прогиб, отклонение
388 деформировать
389 деформация
390 стержень периодического профиля
391 соль для удаления наледи
392 снабжение, распределение
393 нагнетательная труба
394 плотность, объемная масса
395 проект, конструкция, расчет
396 расчетная нагрузка
397 проектная контора
398 проектная величина
399 конструктор

400 испытание до разрушения образца
401 деталь
402 диагональ
403 диаграмма, схема, график
404 диаметр
405 алмазный бур
406 диафрагма, мембрана
407 диафрагмовый насос
408 перемычка
409 дифференциал
410 дифференцировать
411 размер, объем
412 отклонение размера
413 направляющие указания
414 диск, круг
415 непараллельный пучок
416 отсутствие непрерывности, разрезность
417 прерывная гранулометрия
418 демонтировать, распалубить
419 расстояние, отрезок
420 красить, обмазывать
421 распределительная арматура
422 домкрат двойного действия
423 двойной изгиб
424 шпонка, штырь
425 драглайн
426 конструктор, чертежник
427 волочильный станок
428 чертеж, волочение (проволоки)
429 сверло
430 напряжение от вибрации
431 висячая балка
432 высушивать
433 бетон жесткой консистенции
434 сухой док
435 усадка при сушке
436 труба, трубопровод
437 макет, модель, болванка
438 опрокидывающийся кузов
439 дюнный песок
440 прочный, долговечный
441 динамическая нагрузка
442 динамическое испытание
443 динамометр
444 давление грунта
445 земляной склон, наклон грунта
446 сейсмическая нагрузка
447 внецентренная нагрузка
448 эксцентриситет, внецентренность
449 край, ребро
450 угловая балка
451 концевой момент
452 расчетная длина элемента (при расчете на продольный изгиб)
453 эффективная толщина
454 отдача, полезное действие
455 эфлоресценция, выцветание
456 упругий
457 упругая деформация
458 предел упругости
459 упруго опертая балка
460 упругость, эластичность
461 электросварка
462 электротермическое упрочение
463 электротермическое натяжение арматуры
464 повышение, фасад
465 эллипс инерции
466 удлинять, растягивать
467 удлинение
468 эмпирический
469 опалубить, упаковать
470 обшивка, облицовка
471 концевая опора
472 концевой блок
473 арматура концевого блока
474 инженер
475 обволакивать
476 обертывание, покрышка
477 равновесие
478 условие равновесия
479 оборудование
480 возведение, установка, сборка
481 нагрузка при монтаже
482 смета, калькуляция, оценка
483 гравировать, протравлять
484 ванна для травления
485 испытание на травление
486 исследование, осмотр
487 копать
488 экскаватор
489 выполнение
490 тянутый металл
491 расширение
492 температурный шов
493 расширяющийся цемент
494 опыт
495 нагрузка при взрыве
496 вытягивать, удлинять
497 длинномер
498 наружный, внешний
499 наружный пучок
500 внешнее предварительное напряжение

501 наружный пролет
502 наружный вибратор
503 экстраполировать
504 крайнее волокно
505 напряжения в крайних фибрах
506 прессованый экструзированный бетон
507 сварная сетка
508 разрушение
509 теория разрушения
510 предел разрушения
511 разрушающая нагрузка
512 механизм разрушения
513 разрушающий момент
514 стадия разрушения
515 строительные леса
516 веерообразная анкеровка
517 усталость (материалов)
518 сопротивление усталости
519 феррит
520 фибра, волокно
521 фиктивный
522 поясок, буртик
523 мелкий заполнитель
524 мелкий гравий
525 модуль крупности
526 тонкость измельчания, помола
527 отделывать, доводить
528 обработка шаблоном для разравнивания бетонной смеси
529 продолжительность пожара
530 огнестойкий
531 безопасность от огня
532 закреплять
533 момент в заделке
534 момент защемления
535 полка балки
536 ширина фланца, полки
537 барашковая гайка
538 плоская поверхность
539 прочность на изгиб
540 плавучий док
541 нагрузка на перекрытие
542 промывать (водой)
543 путепровод
544 пенобетон
545 клинья, забиваемые навстречу друг другу
546 сила, усилие
547 формула
548 опалубка
549 смазка для опалубки
550 наружный вибратор
551 основание
552 фундаментная плита
553 нагрузка при изломе
554 разрушенный, разбитый
555 каркас
556 частота
557 свежеуложенная (приготовленная) бетонная смесь
558 морозоустойчивый
559 веревочный многоугольник
560 шлакобетон
561 гальванизировать
562 газовая сварка
563 прокладка (изоляция)
564 геометрический
565 балка, балочная ферма
566 балочный мост
567 клей
568 метод склеивания
569 кран большой подъемной силы
570 фракционировать
571 марка, класс, качество
572 гранулометрический состав
573 кривая гранулометрического состава
574 гранитный мелкий щебень, крошка
575 гранулировать
576 гранулометрия
577 графит
578 графостатика
579 гравий
580 гравийный бетон
581 гравийный карман
582 сила тяжести
583 решетка
584 решетчатое перекрытие
585 зажимная анкеровка
586 поверхность грунта
587 грунтовая вода
588 группа стержней
589 жидкий строительный раствор
590 растворомешалка
591 инъецирование раствора
592 воронка для жидкого раствора
593 скважина для нагнетания раствора
594 шланг для раствора
595 машина для заливки раствора
596 насос для подачи раствора
597 насадка для заливки раствора
598 гипс
599 волосная трещина
600 волосяной войлок

601 шпилка; резкий, крутой (о повороте дороги)
602 монтажная арматура
603 твердеть, цементировать
604 закаливать сталь
605 затвердевший цементный камень
606 затвердевание, закалка
607 испытание на твердость
608 твердость, плотность, жесткость
609 пята свода, вута
610 утолщенный бетон (вокруг трубы)
611 внутренняя высота помещения
612 теплота, нагрев
613 теплоизоляция
614 тепловыделение при гидратации
615 тепловыделение при схватывании бетона
616 сталь, подвергнутая термообработке
617 тяжелый бетон
618 высота
619 шланг для высокого давления
620 высотное здание
621 высокопрочный бетон
622 высокопрочная сталь
623 шарнир
624 шарнирная опора
625 выгиб (балки)
626 пористая, сотовая структура
627 крюк
628 спиральная арматура
629 горизонталь
630 горизонтальное сечение
631 лошадиная сила
632 горячая гальванизация (с погружением в жидкость)
633 горячекатаная сталь
634 горячекатаная проволока
635 влажность
636 гидратация
637 усадка при гидратации
638 гидравлическое вяжущее вещество
639 водородная хрупкость
640 гигроскопический
641 гиперболическая оболочка
642 гиотеза
643 ледовая нагрузка
644 мгновенная деформация
645 мгновенная деформация
646 удар
647 ударный коэффициент
648 ударная нагрузка
649 сопротивление удару
650 водонепроницаемый
651 наклон
652 профилированная проволока с вмятинами
653 инерция
654 начальный, первоначальный
655 начальная прочность
656 начальное напряжение
657 вспрыскивать
658 инъекция
659 внутренняя поверхность (стены)
660 внутренний пролет
661 закладывать (деталь в бетон)
662 монолитный бетон
663 наблюдать
664 инспектор, приемщик
665 неустойчивость
666 интегрировать
667 взаимодействия (диаграмма)
668 промежуточная опора
669 внутренний
670 внутренний диаметр
671 внутренний размер
672 внутреннее трение
673 глубинный вибратор
674 интерполировать
675 свободная ширина, пролет в свету
676 исследование
677 заявка на подряд
678 железо
679 железная проволока
680 необратимый
681 изотропный
682 повторять
683 домкрат
684 соединять
685 соединение, стык, шов
686 вибрационный стол
687 поддерживать во влажном состоянии
688 перегиб (кабеля, пучка)
689 лаборатория
690 площадка (лестницы)
691 соединение внахлестку
692 длина нахлестки (перепуска)
693 сварной шов внахлестку
694 боковая поверхность
695 поперечное сокращение
696 решетка
697 решетчатая или сквозная ферма
698 теория решетки
699 выщелачивать

700 свинец
701 тощий бетон
702 длина
703 выравнивать
704 плечо рычага
705 приспособление для подъема
706 подъемное ушко, проушина
707 легкий заполнитель
708 легкий бетон
709 известь
710 выцветание извести
711 известняк
712 предел пропорциональности
713 предельное состояние
714 предельное состояние по раскрытию трещин
715 предельное состояние по трещинообразованию
716 предельное состояние при снижении давления
717 линия действия
718 кривая давления
719 линейный
720 линейная опора
721 нагрузка на погонную единицу
722 перемычка
723 временная нагрузка
724 нагрузка
725 несущая стена
726 несущая способность
727 ковш для нагружения
728 комбинация нагрузок
729 грузовой поезд
730 нагружение
731 запирать, замыкать
732 контргайка
733 стопорный домкрат
734 стопорный плунжер
735 логарифмический
736 стержень продольной арматуры
737 продольная трещина
738 продольная сила
739 продольная балка
740 продольное предварительное напряжение
741 продольная арматура
742 продольное сечение
743 продольный сварной шов
744 длительное испытание
745 петлевая анкеровка
746 потеря напора
747 потеря давления
748 потеря напряжения
749 магнезиальный цемент
750 главная балка
751 главная ферма
752 главная арматура
753 эксплуатация
754 оправка (станка для гнутья стержней)
755 человеко-час
756 рабочая сила
757 изготовлять
758 разметка, маркировка
759 мергель
760 мартенситная закалка с самоотпуском
761 мартенсит
762 материал
763 материал, удерживающийся на сите
764 мера, масштаб, размер
765 измерительная коробка
766 мерный цилиндр
767 механическое испытание
768 механика
769 мембрана
770 напряжение в пластинке
771 меридиан
772 напряжение по меридиану
773 сетка, отверстие сита
774 сетчатая (арматура)
775 метацентр
776 металл
777 металлическая опора
778 микротрещина
779 осевая линия, полоса
780 момент в середине пролета
781 арматура в середине пролета
782 минерад
783 смешивать
784 смесь
785 смесительный барабан
786 завод или установка для приготовления бетона
787 шнек, винт для смешивания бетона
788 вода затворения
789 испытание на моделях
790 метод расчета по допускаемым напряжениям (по отношению модулей упругости)
791 модуль
792 модуль упругости
793 влажность
794 момент
795 коэффициент момента
796 эпюра моментов

797 момент инерции
798 монолитный
799 монорельс
800 (известковый) строительный раствор
801 формовать
802 форма
803 подвижная нагрузка
804 безбалочное перекрытие
805 игла Вика (для определения схватывания цемента)
806 нейтральная ось
807 испытание неразрушающими методами
808 эпюра нормальных сил
809 нормальное напряжение
810 пильчатая или шедовая крыша
811 жесткая бетонная смесь
812 надрез, концентратор напряжений
813 влияние надреза
814 коэффициент ударного сопротивления
815 численный, цифровой
816 гайка
817 гаечное закрепление
818 косой, наклонный
819 масло, жидкая смазка
820 инструкция, заказ
821 прямоугольный
822 ортотропный
823 колебаться
824 арматура с овальным поперечным сечением
825 проволока с овальным поперечным сечением
826 перегружать
827 путепровод над дорогой
828 ацетиленовая сварка
829 замонолитить (горизонтальный шов)
830 набивочное кольцо
831 панель
832 парабола
833 параметр
834 часть объема
835 часть веса
836 частичный, неполный
837 неполное предварительное напряжение (с допущением растяжения в бетоне при эксплуатационных нагрузках)
838 размер частицы
839 гранулометрический анализ
840 прочность частицы
841 патент
842 патентировать (сталь)
843 дорожное покрытие
844 горошковый гравий (размер 6–19 мм)
845 перлит
846 проникновение воды
847 пенетрометр
848 процентное содержание по объему
849 процентное содержание по весу
850 допускаемое напряжение
851 испытание методом фотоупругости
852 бык, пристань
853 свая
854 оголовок сваи
855 ствол сваи
856 пята сваи
857 труба
858 шаг винтовой нарезки или шаг спирали
859 устанавливать, укладывать бетон
860 чистый, неармированный
861 гладкий стержень
862 гладкая проволока
863 плоскость
864 штукатурить
865 пластмасса
866 пластическая деформация
867 пластическая усадка
868 пластическая деформация
869 пластификатор
870 пластичность (поведение материалов)
871 пластичность (консистенция)
872 плита
873 работа плиты
874 зазор
875 пробка
876 отвес
877 плунжер
878 поршневый насос
879 точка нулевого момента
880 полировать
881 отшлифованная поверхность
882 заселение, популяция
883 пористый
884 портальный кран
885 портальная рама
886 положение
887 натяжение арматуры на бетон
888 срок сохранности
889 бетонировать, укладывать бетон

890 механическая лопата
891 изготавливать на заводе, полигоне
892 сборный бетон (железобетон)
893 балка "Префлекс"
894 предварительный расчет
895 предварительное испытание
896 сохранять
897 датчик давления
898 манометр
899 подвергать предварительному напряжению
900 предварительное напряжение
901 предварительно напряженный
902 стенд для натяжения арматуры (длинный)
903 стенд для натяжения арматуры (короткий)
904 предварительное напряжение путем навивки (метод карусели)
905 усилие предварительного напряжения
906 арматура для предварительного напряжения
907 система предварительного напряжения
908 натяжение арматуры на упоры
909 главное напряжение
910 главное растягивающее напряжение
911 призматический
912 теория вероятности
913 поточная линия
914 профилирование
915 подпирать, подкреплять
916 подпорка
917 дозировать
918 выпуск арматуры
919 пульсирующее напряжение
920 пемзобетон
921 накачивать
922 пробивать, штамповать
923 напряжение при штамповании
924 обрешетина
925 оттяжка вниз (напрягаемой арматуры)
926 кварцевая пыль
927 конструкция набережной
928 стенка набережной
929 радиальный
930 радиальное напряжение
931 радиус кривизны
932 радиус инерции
933 дождевая вода
934 наклонная свая
935 наклонное положение
936 пандус, откос насыпи
937 выборочное испытание
938 быстротвердеющий цемент
939 скорость возрастания нагрузки
940 готовая смесь
941 завод товарного бетона
942 ниша, паз, углубление
943 уменьшение, сокращение
944 барабан, катушка
945 отказ (сваи при забивке)
946 инструкция
947 армировать (бетон), усиливать
948 жедезобетон
949 арматура
950 каркас арматуры
951 арматурные ножницы
952 арматурный детектор
953 процент армирования
954 арматурный стержень
955 арматурная сталь
956 относительная влажность
957 релаксация
958 освобождать, отпускать
959 отпускать (напряжение)
960 ослаблять (клин)
961 разгружать
962 исследование; исследовать
963 резервуар
964 упругое соединение
965 сопротивление
966 сварка встык контактная
967 сварка методом сопротивления
968 устойчивый
969 равнодействующий
970 подпорная стенка
971 замедлять
972 напрягать повторно
973 испытание на загиб
974 обратимый
975 ребро
976 ребристое перекрытие
977 ребристая плита
978 жирный подстилающий слой (в качестве первой заливки в опалубке)
979 жирный бетон
980 жесткий узел
981 жесткость
982 кольцевая балка
983 стрела арки, подъем арки
984 речной гравий
985 речной песок

986 анкер в скале
987 балансир
988 качающаяся опора
989 ручная тачка с опрокидывателем (для бетона)
990 качаться
991 ролик
992 цилиндрическая подвижная опора
993 катанка
994 волнистое покрытие
995 крыша, покрытие
996 кровельная балка
997 стропильная система покрытия
998 нагрузка на покрытие
999 конструкция покрытия
1000 вращение
1001 круглый стержень
1002 сито с круглыми отверстиями
1003 гравий, галька
1004 резиновый подшипник, опора
1005 резиновая прокладка, подушка
1006 взлетно-посадочная дорожка
1007 коррозия, ржавчина
1008 подушка, седло, башмак
1009 безопасность, надежность
1010 трещиностойкость; надежность против появления трещин
1011 безопасность от разрушения
1012 обазец
1013 песок
1014 резервуар для песка
1015 обрабатывать струей песка
1016 многослойная плита
1017 насыщенный
1018 леса, подмостки
1019 масштаб (размер)
1020 отделяться слоями
1021 рассеивать
1022 проект, схема, план
1023 размывать, подмывать
1024 шаблон для разравнивания бетонной смеси
1025 сито
1026 номер сита
1027 закрепление анкерным болтом
1028 винтовой домкрат
1029 винтовая нарезка, резьба
1030 муфта с резьбой
1031 разметчик
1032 очищенная щеткой поверхность
1033 морской гравий
1034 закрывать, запечатывать
1035 состав, предохраняющий бетон от высыхания
1036 дополнительное напряжение
1037 поперечное сечение
1038 момент сопротивления (сечения)
1039 профильная сталь
1040 расслоение
1041 размечать (размеры)
1042 установочный винт
1043 схватывание
1044 ускоритель схватывания
1045 замедлитель схватывания
1046 осадка (грунта)
1047 канализационная труба
1048 вибрационное сито, грохот
1049 дробить
1050 поперечная сила
1051 эпюра поперечной силы
1052 модуль сдвига
1053 поперечная арматура
1054 срезывающее напряжение
1055 прочность на срез
1056 срез
1057 обшивка (кабеля)
1058 листовой материал (пластик, металл)
1059 тонкое листовое железо
1060 шпунтовая свая
1061 шпунтовая стенка
1062 оболочка (конструктивная)
1063 крепь, подкос
1064 подпорка
1065 сокращать
1066 кратковременное испытание
1067 торкрет бетон
1068 усадка
1069 усадочные трещины
1070 коэффициент усадки
1071 градиент усадки
1072 усадочный шов
1073 модуль усадки
1074 противоусадочное армирование
1075 усадочное напряжение
1076 устанавливать опалубку
1077 сито
1078 ситовой анализ
1079 остаток на сите
1080 бункер, силос
1081 эффект силоса
1082 спекание, агломерация
1083 стройплощада
1084 бетон, приготовленный на стройке
1085 косой

1086 косая плита
1087 косой мост, перекрытый плитой
1088 квалифицированный рабочий
1089 скип, вагонетка
1090 плита
1091 плитный мост
1092 плитное перекрытие
1093 провес, люфт
1094 шифер
1095 рукав
1096 гибкость
1097 скользить
1098 вдвигать одно в другое
1099 логарифмическая линейка
1100 подвижная опалубка
1101 реакция скольжения
1102 скольжение, сдвиг
1103 эллинг, слип, стапель
1104 наклон, уклон
1105 угол, наклон
1106 паз
1107 медленно схватывающийся цемент
1108 шлюз
1109 осадка конуса
1110 конус для измерения осадки бетонной смеси
1111 определение осадки конуса
1112 снеговая нагрузка
1113 твердый
1114 сплошной
1115 балка со сплошной стенкой
1116 прозвучивать
1117 неизменность
1118 сплошность, плотность
1119 распорка
1120 маячный блок
1121 распорное кольцо
1122 раскалывание
1123 пролет
1124 направление пролетов
1125 удельный
1126 удельный вес
1127 удельная теплота
1128 правила и условия контракта
1129 сферическая оболочка
1130 сферическое напряжение
1131 спиральная арматура
1132 спиральная пружина
1133 спирально армированная колонна
1134 стыковая накладка
1135 петлевой перехлест
1136 патрубок сращивания
1137 раскалывать
1138 раскалывающие напряжения
1139 катушка
1140 точечная сварка
1141 распределенная (веерная) анкеровка
1142 упругая постоянная
1143 стержень квадратного сечения
1144 устойчивость
1145 стабилизировать
1146 складывать штабелем
1147 стадия
1148 норма; типовой стандарт
1149 стандартное отклонение
1150 стандартная прочность
1151 стандартизация
1152 стандартизировать
1153 напряженное состояние
1154 статическая нагрузка
1155 статический момент
1156 статически определимый
1157 статически неопределимый
1158 статика
1159 статистика
1160 ванта
1161 пропаривать; пар
1162 пропаривание (бетона)
1163 сталь, арматура
1164 стальная плита
1165 стальной лист
1166 стальная проволока
1167 стальной проволочный канат
1168 арматурщик
1169 клещи арматурные
1170 бетон жесткой консистенции
1171 придавать жесткость
1172 жесткость
1173 скоба, хомут
1174 шаг хомутов
1175 мелкий щебень, крошка
1176 склад
1177 этаж
1178 деформация
1179 тензометр
1180 прядь, канат
1181 прочность, сила
1182 прочность материалов
1183 прочность при длительной нагрузке
1184 напряжение
1185 круги напряжений (круги Мора)
1186 концентрация напряжений
1187 коррозия под напряжением
1188 понижение напряжения

1189 распределение напряжений
1190 градиент напряжения
1191 скачок напряжения
1192 потеря напряжения
1193 участок напряжения, напряженная зона
1194 кривая напряжение-деформация
1195 траектории напряжений
1196 перемена напряжений
1197 волна напряжения
1198 мост с напряжением по участкам
1199 упор для создания напряжения
1200 оборудование для осуществления напряжения
1201 головка для натяжения арматуры
1202 домкрат для натяжения арматуры
1203 оператор, осуществляющий натяжение арматуры
1204 схема напряжения
1205 растягивать
1206 снимать опалубку
1207 срезать излишек
1208 конструктивный элемент
1209 запас конструктивной прочности
1210 сооружение, конструкция
1211 подстилающий грунт
1212 фундамент, нижнее строение
1213 подземный
1214 налагать
1215 сульфатный цемент
1216 надзор
1217 контролер
1218 опора, станина
1219 опорный момент
1220 реакция опоры
1221 поверхность
1222 поверхностная арматура
1223 поверхностное натяжение
1224 поверхностная обработка
1225 поверхностный вибратор
1226 висячее покрытие
1227 висячий мост
1228 разбухать
1229 синтетическая смола
1230 прихватка
1231 трамбовать, наполнять
1232 трамбованный бетон
1233 тангенс
1234 точка касания
1235 тангенциальный
1236 резервуар
1237 заострять к конпу
1238 эпоксидная смола
1239 отпускать
1240 температурный градиент
1241 температурная усадка
1242 температурное напряжение
1243 напрягаемая арматура
1244 предварительное напряжение без натяжения арматуры
1245 растягивающая сила
1246 деформация при растяжении
1247 предел прочности на растяжение
1248 растягивающее напряжение
1249 испытание на растяжение
1250 зона растяжения
1251 растягивать
1252 растяжение
1253 растянутая свая
1254 растянутая арматура
1255 оборудование для натяжения
1256 домкрат для натяжения
1257 испытание, опыт, проба, проверка
1258 испытываемый стержень арматуры
1259 опытная балка
1260 образец-кубик (бетона)
1261 образец-цилиндр (бетона)
1262 пробная нагрузка
1263 испытываемый образец
1264 испытания
1265 теоретическая механика
1266 толщина
1267 триксотропный
1268 трехшарнирная рама
1269 распор
1270 арка с уравновешивающим распором
1271 стяжной хомут
1272 связь (арматуры)
1273 шпала
1274 связь, стяжка
1275 проволочная связь
1276 качающийся барабан (смесителя)
1277 допуск
1278 инструменты
1279 сцепление, зуб, зубец
1280 верхний стержень
1281 верхнее волокно
1282 верхняя полка (балки)
1283 верхняя сетка
1284 верхняя часть
1285 гаечный ключ
1286 крутящий момент
1287 напряжение при кручении
1288 грузонапряженность дороги
1289 зона передачи, анкеровки

1290 поперечина
1291 поперечная трещина
1292 поперечный элемент
1293 предварительное напряжение, поперечное
1294 поперечная арматура
1295 трасс, тонкий вулканический туф
1296 трассовый портландцемент
1297 ход, перемещение (домкрата)
1298 воронка для бетонирования под водой
1299 подводное бетонирование
1300 эстакадная рама
1301 трехосное предварительное напряжение
1302 трехосное напряжение
1303 лопатка, кельма, мастерок
1304 автобетономешалка
1305 туннель
1306 стяжная муфта
1307 скручивать
1308 U-образный крюк
1309 предел прочности на сжатие
1310 метод расчета по разрушающим усилиям
1311 разрушающий момент
1312 предел прочности
1313 критческое напряжение
1314 предел прочности на растяжение
1315 арматурный элемент без сцепления с бетоном
1316 разматывать
1317 подземный
1318 подводный бетон
1319 разбирать, демонтировать
1320 равномерное нагружение
1321 равномерно распределенная нагрузка
1322 единица, комплект, элемент
1323 развинчивать
1324 неквалифицированный рабочий; неквалифицированный работник
1325 ненапряженный
1326 ненапрягаемая арматура
1327 разматывать
1328 высаживать
1329 предел деформации при высадке
1330 вакуум-бетон
1331 переменный
1332 свод
1333 вектор
1334 патрубок
1335 выпускной стояк
1336 вертикальный
1337 вертикальный разрез
1338 виадук
1339 вибрировать
1340 вибрационная (и отделочная) машина
1341 вибрационная балка, вибро-рейка
1342 вибрационный стол
1343 вибрация
1344 вибратор
1345 вибрационное сито; грохот
1346 ферма Виренделя
1347 вязкость
1348 пустоты, поры
1349 содержение пустот
1350 объем пустот
1351 объем
1352 дозирование по объему
1353 кессонное перекрытие
1354 схватка, насадка
1355 стена
1356 косой, наклонный, скрученный
1357 скручивание, перекашивание
1358 шайба
1359 водо-цементное отношение
1360 водосодержание
1361 запорный кран, гидроизоляция
1362 водоснабжение
1363 водонапорная башня
1364 водонепроницаемый
1365 водонепроницаемый бетон
1366 водонепроницаемость
1367 износ
1368 нормы климатических условий
1369 устойчивый против атмосферных влияний
1370 стенка (балки)
1371 клин
1372 анкеровка клиньями
1373 домкрат для заклинивания
1374 взвешивать, весить
1375 дозировка по весу
1376 весовой дозатор
1377 весы
1378 вес, груз, масса
1379 водосливная плотина
1380 сваривать
1381 сварной шов
1382 длина сварочного щва
1383 свариваемый
1384 оборудование для сварки
1385 белый цемент
1386 белый песок

1387 ширина
1388 ветер
1389 ветровая связь
1390 ветровая нагрузка
1391 давление ветра
1392 обмоточная машина
1393 барашковая (крыльчатая) гайка
1394 проволока, трос
1395 проволочное закрепление
1396 проволочный зажим
1397 моток проволоки
1398 барабан для наматывания проволоки
1399 проволочное сито
1400 машина для намотки проволоки (для резервуаров)
1401 трение, возникающее от случайного отклонения каналов
1402 древесина (лесоматериал)
1403 деревянная щепа
1404 благоприятный для работы (погода)
1405 удобоукладываемость
1406 рабочий чертеж
1407 рабочее напряжение
1408 рабочий, работник
1409 мастерская, цех
1410 шнековый, винтовой транспортер
1411 обертывать, обматывать
1412 исследование рентгеновскими лучами
1413 площадка текучести
1414 предел текучести
1415 предел текучести
1416 текучесть (стали)

РУССКОЕ АЛФАБИТНОЕ ОГЛАВЛЕНИЕ
RUSSIAN ALPHABETICAL INDEX

А а

Б б

В в

Н н

О о

П п

Р р

С с

Т т

У у

Ф ф

Х х

Ц ц

Ч ч

Ш ш

Щ щ

Э э

Я я